Liquid Silicone Rubber

Scrivener Publishing
100 Cummings Center, Suite 541J
Beverly, MA 01915-6106

Publishers at Scrivener
Martin Scrivener (martin@scrivenerpublishing.com)
Phillip Carmical (pcarmical@scrivenerpublishing.com)

Liquid Silicone Rubber

Chemistry, Materials, and Processing

Johannes Karl Fink

Scrivener Publishing

WILEY

This edition first published 2019 by John Wiley & Sons, Inc., 111 River Street, Hoboken, NJ 07030, USA and Scrivener Publishing LLC, 100 Cummings Center, Suite 541J, Beverly, MA 01915, USA

For more information about Scrivener publications please visit www.scrivenerpublishing.com.

Wiley Global Headquarters
111 River Street, Hoboken, NJ 07030, USA

For details of our global editorial offices, customer services, and more information about Wiley products visit us at www.wiley.com.

Library of Congress Cataloging-in-Publication Data

ISBN 978-1-119-63133-0

Cover image: Pixabay.Com
Cover design by Russell Richardson

Set in size of 11pt and Minion Pro by Manila Typesetting Company, Makati, Philippines

10 9 8 7 6 5 4 3 2 1

Contents

Preface

The scientific literature with respect to liquid functional rubbers is collected in this monograph. The text focuses on the basic issues and also the literature of the past decade. The book provides a broad overview of the materials used therein.

In particular, materials and compositions for liquid functional rubbers are discussed. Also, methods of curing and special properties are described, such as tracking and erosion resistance, adhesion properties, storage and thermal stability. Methods of curing are precision casting, hybrid additive manufacturing, peroxide curing, ultraviolet curing, liquid injection molding, or hot embossing. Special fields of applications are automotive and underwater applications, electrical and optical uses, and medical uses.

The text is usable for a lot of different audiences, such as education in university, but also for researchers and practitioners in the fields of electrical engineers, automotive engineers and researchers in medical fields.

How to Use This Book

Utmost care has been taken to present reliable data. Because of the vast variety of material presented here, however, the text cannot be complete in all aspects, and it is recommended that the reader study the original literature for more complete information.

Index

There are three indices: an index of acronyms, an index of chemicals, and a general index. In the index of chemicals, compounds that occur extensively are not included at every occurrence, but rather when they appear in an important context. When a compound is found in a figure, the entry is marked in boldface letters in the chemical index.

Acknowledgements

I am indebted to our university librarians, Dr. Christian Hasenhüttl, Gerlinde Iby, Margit Keshmiri, Friedrich Scheer, Christian Slamenik, Renate Tschabuschnig, and Elisabeth Groß for their support in the literature acquisition. In addition, many thanks to the head of my department, Professor Wolfgang Kern, for his interest and permission to prepare this text.

I also want to express my gratitude to all the scientists who have carefully published their results concerning the topics dealt with herein. This book could not have been otherwise compiled.

Last, but not least, I want to thank the publisher, Martin Scrivener, for his abiding interest and help in the preparation of the text. In addition, my thanks go to Jean Markovic, who made the final copyedit with utmost care.

Johannes Fink
Leoben, 16th May 2019

1

Materials

1.1 History

The history of silicone rubber, its development and technological progress has been described (1–3). Its first commercial development was in 1944. Pioneering research on silicon opened the door to the development of silicone polymers and silicone rubber. The substitution of two methyl groups on silicon was present in the first examples of silicone rubber and still is the predominant organic group in commercial silicone rubber today (1).

Silicone rubbers have filled a need in the marketplace because of their combination of unusual properties not found in other rubbers. The alternating inorganic main-chain atoms of silicon and oxygen, and the two pendant organic groups, primarily methyl, provide strong chain bonds, backbone flexibility, ease of side-group rotation, and low inter- and intramolecular forces (1).

Also, industrial organosilicone materials, their environmental entry and their predicted fate have been described in a monograph (4). Furthermore, a comprehensive overview of the issues of organosilicon compounds has been presented. Commercial products, such as sealants, adhesives, and coatings are reported (5, 6).

1.2 Properties

1.2.1 Tracking and Erosion Resistance

Silicon rubber (SR) also offers excellent electrical performance under contaminated environments. However, a pristine silicone rubber has low thermal properties and this may cause tracking and erosion failure due to severe dry band arcing and ohmic heating on the insulating surface (7).

The effect of material thermal characteristics on the tracking and erosion resistance of silicone rubber filled with micron-sized alumina trihydrate, aluminum nitride and boron nitride particles was investigated (7). Composites with different loading were synthesized by dispersing the particles in pristine room temperature vulcanized (RTV) silicone rubber and an IEC 60587 inclined plane test (8) was conducted to evaluate tracking and erosion resistance. Apart from physical parameters and leakage current, an infrared thermal imager was used to measure the surface temperature distribution during the course of the inclined plane test. Experimental results showed tracking and erosion resistance is significantly enhanced with addition of boron nitride particles followed by alumina trihydrate.

Aluminum nitride composites exhibit a poor tracking and erosion resistance, similar to pristine silicone rubber. It has been concluded that the addition of boron nitride composites improves the ability to impede the tracking and erosion process, the reasons being better thermal stability and enhanced thermal conduction in the discharge region. On the other hand, infrared analysis revealed that thermal accumulation is remarkably higher in aluminum nitride composites, which promotes dry band arcing and results in tracking and erosion failure (7).

To improve the tracking and erosion resistance performance of addition-cure liquid silicone rubber (LSR) without alumina trihydrate, addition-cure LSR samples with different fumed silica mass fraction and platinum catalyst concentration were prepared (9).

The tracking and erosion resistance performance of the samples were evaluated. Also, the thermal decomposition products were detected. Based on the comparative analysis of thermal decomposition characteristics, the morphology, crystal structure, and improvement mechanism were assessed.

The results of this study showed that both increasing the concentration of the platinum catalyst and the mass fraction of fumed silica can improve the tracking and erosion resistance performance, whereas the former was more effective. Further analysis indicated that the platinum catalyst played a very important role in the thermal decomposition process of addition-cure LSR. Increasing the platinum catalyst content can not only promote the crosslinking reaction between the methyl groups, but also can suppress backbone decomposition (9).

1.2.1.1 *Impact of Various Fillers*

The influence that various fillers having different sizes, from 0.3 μm to 18 μm, and surface modifications (unmodified, modified by the material supplier and *in-situ* modified during compounding) have on the erosion resistance of high temperature crosslinked silicone rubber composites was analyzed. The particles used were aluminium trihydrate, alumina (Al_2O_3) and silica (SiO_2). The main focus was on aluminium trihydrate fillers since they have the ability to release water at elevated temperature.

By a simple water storage test under defined conditions the water uptake for different composites was analyzed to assess the effectiveness of an *in-situ* modification. The inclined plane test according to IEC 60587 (8), and the high voltage, low current dry arc test according to IEC 61621 (10), were used to determine the erosion resistance of the different samples.

For the inclined plane test, an adapted evaluation model was applied. This test is known to have a wide scatter in the case of material formulations, which are on the borderline of passing or failing the test and therefore show substantial erosions on certain samples. The scatter could be reduced by evaluating the eroded volume by using samples only, which showed a limited erosion length. It was found for aluminium trihydrate, that larger particles show slightly better results than smaller particles. This can be explained by the formation of boehmite AlO(OH) for the larger particles, which causes a release of the bound water over a wider temperature range.

This effect could be confirmed by thermogravimetric analysis (TGA). The surface modification of the particles with vinyltrimethoxysilane and methyltrimethoxysilane, c.f. Figure 1.1, did not improve the erosion resistance significantly, but reduces the water uptake to a large extent, which is advantageous for the retention of the hydrophobicity. In order to achieve a low erosion rate, high filler loadings are essential (11).

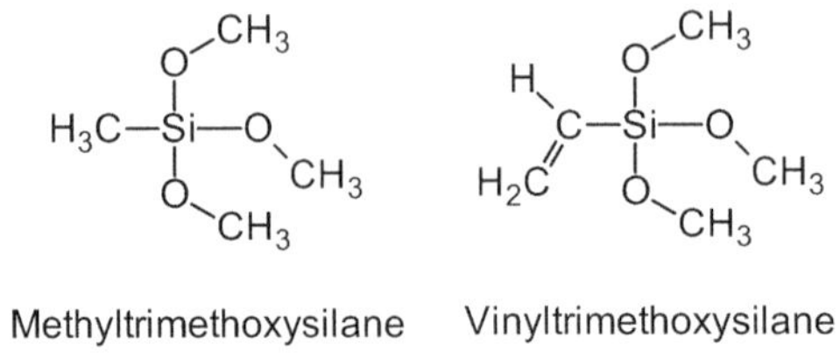

Figure 1.1 Methyltrimethoxysilane and Vinyltrimethoxysilane.

1.2.2 *Enhancing Strength*

Silica is the most widely used filler to reinforce LSR, but the high viscosity of a LSR/silica suspension significantly limits its processing flexibility (12).

To balance the processability and the reinforcing efficiency of LSR/silica systems, two kinds of enols, i.e., propenol and 1-undecylenyl alcohol, and a saturated alcohol, 1-undecylic alcohol, were employed to modify the silica surface. The compounds used here are shown in Figure 1.2.

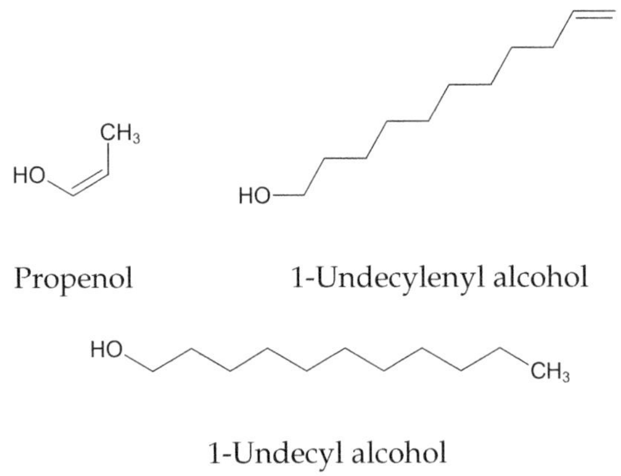

Figure 1.2 Alcohols.

Modified silica samples were prepared via surface esterification of silica with alcohols by the reflux method. The calculated amount of precipitated silica (Rhodia Z-142), xylene, and alcohol were added into a reaction flask. Then a solid-liquid extractor containing sufficient CaH_2 and a condenser with a $CaCl_2$ drying tube were mounted on the reaction flask. The mixture was heated to 160°C and refluxed for 24 *h*. Finally, the samples were filtered and washed with ethyl acetate at least five times and dried in a vacuum oven at 120°C for 8 *h*.

The carbon contents, alkoxyl surface densities, and bound rubber contents of various silica samples are shown in Table 1.1.

Table 1.1 Properties of various silica samples (12).

Modifier	Carbon content /[%]	Alkoxyl surface density /[$\times 10^6 mol\, m^{-2}$]	Bound rubber content /[%]
None	0.08	—	0.81
1-Undecylic alcohol	4.12	1.73	0.52
1-Undecylenyl alcohol	4.80	2.02	0.54
Propenol	1.40	2.16	0.59

Obviously, the bound rubber content is largely reduced after silica surface modifications, suggesting that silica poly(dimethyl siloxane) (PDMS) interaction is effectively lowered by silica surface modification. Comparing different modifiers, there is not much difference between 1-undecylic alcohol modified silica and 1-undecylenyl alcohol modified silica, while bound rubber content for propenol modified silica is a little higher. It seems that modifiers with higher carbon chain lengths are more effective in lowering the PDMS silica interactions. Though propenol has a little higher density on the silica surface, it is the combined effect of grafting density and modifier chain length that determines the effect of silica surface modifications.

Various rheological tests were carried out to investigate the processability as well as filler networking and the crosslinking processes of the modified systems.

Polymer/filler composites are known to exhibit a complex rheological behavior that reflects interactions among components in the system. In PDMS/hydrophilic silica suspensions, there are strong interactions between PDMS and silica and among silica, mainly via hydrogen bonding, which would cause a drastic increase in viscosity or reduction in the mobility of PDMS, consequently deteriorating the processing properties of LSR.

The effect of silica surface modification on viscosity of the suspension is believed to be determined by the combined effects of the type of grafting enol (saturated or unsaturated), grafting carbon content (or carbon chain length), and grafting density

Tensile tests were also adopted to verify the reinforcing effect. Tensile strength of various crosslinked LSR systems as a function of silica content are shown in Figure 1.3. Also, the tensile elongation break of various crosslinked LSR systems as a function of silica content are shown in Figure 1.4.

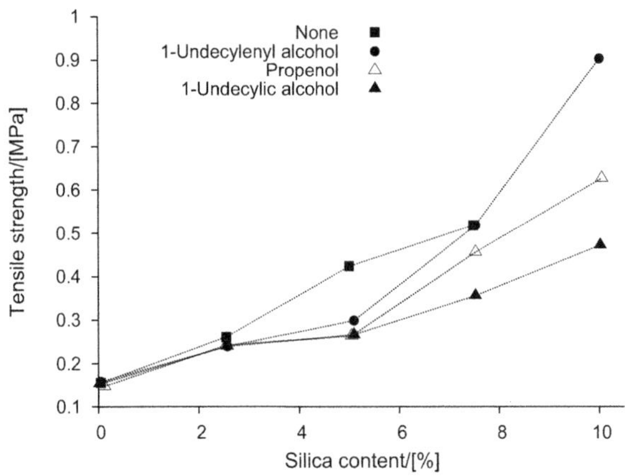

Figure 1.3 Tensile strength of various crosslinked LSR systems as a function of silica content (12).

The investigated systems follow a similar trend: First, the elongation increases gradually with the content of silica, and then decreases because the rigidity of the filler begins to dominate the property. Comparing the systems reinforced with different kinds of silica at the same silica content, their orders of elongation at break (from high to low) are listed as follows: 1-Undecylenyl alcohol modified system, untreated silica reinforced system, propenol-modified system, and 1-undecylic alcohol modified system (elongations of the last two systems are very close). For instance, at the silica content of 2.5%, where tensile strengths of various systems are similar, the average elongations of the reinforced rubbers at the above sequence are 268%, 249%, 231%, and 225%, respectively.

In summary, it was found that surface modification of silica by 1-undecylenyl alcohol could significantly reduce the viscosity of its

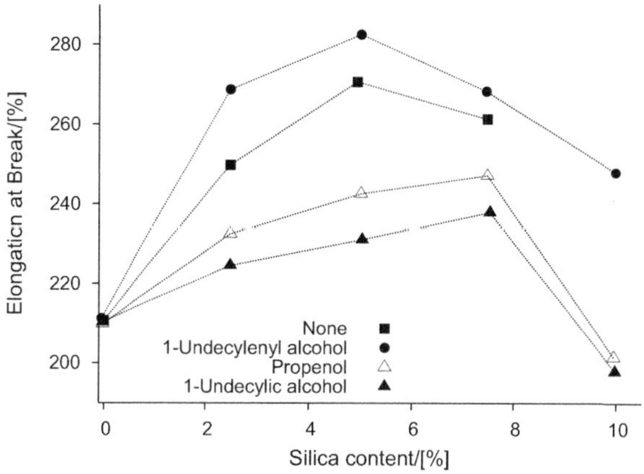

Figure 1.4 Tensile elongation break of various crosslinked LSR systems as a function of silica content (12).

suspension with LSR. Also, the mechanical strength of LSR could be largely enhanced by six times with 10% modified silica (12).

1.2.3 *Surface Treatment*

1.2.3.1 *Prevention of Microwrinkles*

Attempts have been made to prepare a homogeneous film on a PDMS surface and to eliminate microwrinkles from the surface (13). Hydroxy groups were generated on a PDMS surface using different methods.

Because the hydroxylation process changes the chemical composition of the PDMS surface, resulting in a cracked surface, the selection of the best method for surface treatment with minimized surface microwrinkles and cracks was tried.

The results obtained from scanning electron microscopy showed that using the pulsed ultraviolet-ozone radiation method with a controlled duration time, ozone treatment, continuous ultraviolet-ozone treatment using a glass filter, and water media in ultraviolet-ozone treatment was more effective than other methods evaluated in the study to prevent microwrinkles.

Also, the results obtained from contact angle measurements and attenuated total reflectance Fourier transform infrared (FTIR) spectroscopy revealed that the ultraviolet-ozone treatment in the presence of a water medium created more hydroxy groups in comparison to other methods (13).

1.2.3.2 Antifouling Surfaces

Biomimetic Antifouling Surface. Despite the distinct advantages of PDMS for biomedical applications, because of its hydrophobic nature, this material suffers from non-specific protein adsorption and platelet adhesion and activation when used as a blood-contacting material (14).

To confer hydrophilicity and biomolecules repelling characteristics, well-defined and high-density poly(2-hydroxyethyl methacrylate) brushes were synthesized via surface-initiated atom transfer radical polymerization (SI-ATRP) on the PDMS substrate (14).

First, the PDMS surface was activated using an ultraviolet/ozone wet treatment in water media to introduce hydroxy moieties without scarifying the surface property, resulting in a crack-free SiO_2 surface. Then, 3-(2-bromoisobutyramido)propyl(trimethoxy)silane, c.f. Figure 1.5, the active atom transfer radical polymerization initiator, was immobilized on the ultraviolet/ozone-treated PDMS surface to prepare a thin layer of hydrophilic poly(2-hydroxyethyl methacrylate) brush on PDMS substrate, exhibiting excellent protein and platelet resistance. Poly(2-hydroxyethyl methacrylate) brushes supply a biomimetic feature by combining antifouling properties due to hydrophilic characteristic with bioactive properties resulting from the presence of high density hydroxy groups, which can be subsequently used for the conjugation of biomolecules.

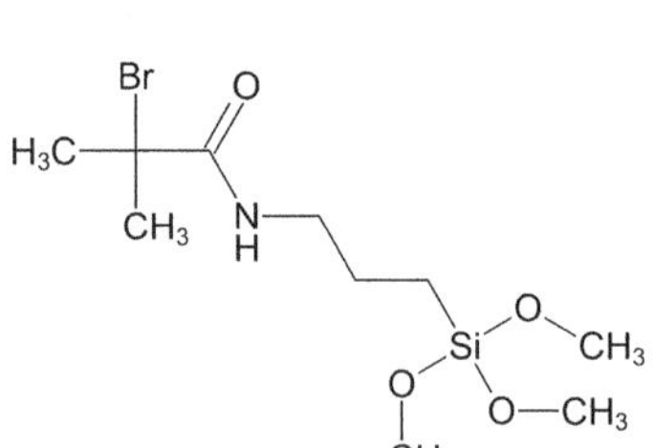

Figure 1.5 3-(2-Bromoisobutyramido)propyl(trimethoxy)silane.

The results of the study indicated that grafting of poly(2-hydroxyethyl methacrylate) chains on the PDMS surface enhances the surface wettability, which leads to a decrease in non-specific protein adsorption and platelet adhesion compared to the bare PDMS substrate. The adhered platelets on the poly(2-hydroxyethyl methacrylate)-tethered PDMS substrate maintain their normal round morphology. In addition, the conjugated gelatin macromolecules on the tethered poly(2-hydroxyethyl methacrylate) chains promote the adherence and growth of human umbilical vein endothelial cells via ligand-receptor interactions (14).

Antifouling Marine Paints. Antifouling marine paints are topcoats intended to prevent the attachment of animals or plants to the lower parts of the hulls of ships (15). They are used for reasons of safety, maintaining the maneuverability of ships, reducing fuel consumption, combating corrosion and weighing-down of structures.

Biofouling is a major problem resulting from the immersion of materials in marine environments. The prevention of this phenomenon represents a considerable maintenance cost. Specifically, the formation of biofouling occurs during immersion in seawater, where a layer of organic and inorganic molecules is adsorbed to the surface of the material extremely rapidly. This layer of adsorbed material, or biofilm, serves as a mediator for the adhesion of the bacteria present in suspension in the marine environment.

This colonization of the surface by marine bacteria is rapid and a stationary state is reached after a period of a few hours to a few days. Finally, other marine organisms colonize the surface, the adherent bacteria recruiting these other organisms. All these live organisms attached to the surface constitute the biofouling or fouling. The adhesion of marine fouling concerns any structure immersed in the sea: ships, pipelines, cooling towers and circuits, harbor structures, marine sensors, aquaculture systems, etc. The damage caused is considerable and diverse. Specifically, the structures become coated, for example, with organisms which have a negative effect on the performance levels of the structures. In particular, for the hulls of ships, the incrustation of various marine organisms increases the friction between the hulls of the ships and the seawater, which reduces the speed and can lead to greater fuel consumption (15).

Thus, the bottom of a ship which is not protected by an antifouling system can, after less than six months spent at sea, be covered with

150 *kg* of fouling per square meter. In order to avoid economic loss, and also in order to more successfully inhibit corrosion phenomena, antifouling paints, the objective of which is to prevent or notably reduce the soiling due to the incrustations of marine organisms, are applied to the immersed parts of the structures exposed to water.

The principle of antifouling paints is based on the controlled release of the active substance at the interface between the surface and the seawater. The effectiveness of the paint is maintained as long as the concentration of active substance released at the surface is effective and regular. Most antifouling paints therefore contain a biocidal product which is most commonly an organometallic compound (based on tin, copper or zinc) or an organic compound (fungicide, algicide, bactericide) which prevents adhesion of the marine soiling owing to the toxic activity thereof (15).

Tributyltin, which is very effective, was therefore the biocide most commonly used in antifouling paints, but this product, its degradation molecules and its metabolites proved to be seriously and sustainably polluting. For these reasons, the International Maritime Organization prohibited the use of tin-based antifouling paints. The antifouling paints used today are mainly based on copper-containing compounds and/or on synthetic chemical compounds, but also are based on silicone polymers. With regard to the copper-based paints, although they are less toxic than tin salts, they are virtually always formulated with a massive proportion of cuprous oxide. However, they are effective only against the marine fauna, and, in order to combat the growth of algae, it is essential to add herbicides, which can pose new threats to the environment.

These silicone-based paints forming an antifouling coating are very innovative:

1. They are completely friendly to the marine environment: no metal waste, and
2. They improve the glide of ships, thus reducing by 1 to 5% their fuel consumption and therefore their greenhouse gas emissions.

An antifouling topcoat based on a silicone elastomer can contain fluids which improve the antifouling effect, in particular (15):

1. Methylphenylpolysiloxane oils (16),

2. A hydrocarbon-based liquid compound, for example a polyolefin,
3. A plasticizer,
4. A lubricating oil,
5. Liquid paraffins and waxy masses of the petrolatum type,
6. A thermoplastic polymer such as poly(vinyl chloride),
7. A vinyl chloride/vinyl acetate copolymer, or
8. Cationic, anionic, nonionic or amphoteric surfactants.

In order to form the silicone elastomer coating, the silicone formulations generally used involve a silicone oil, generally a reactive poly(dimethyl siloxane) with hydroxylated endings, which optionally prefunctionalize with a silane so as to have alkoxy ends, a crosslinking agent and a polycondensation catalyst, conventionally a tin salt or an alkyl titanate, a reinforcing filler and other optional additives such as bulking fillers, adhesion promoters, and dyes (15).

The catalysts which are used are based on tin, titanium, or amine or compositions. Catalysts based on tin and on titanium are very effective (17).

1.2.3.3 Bilayers for Gas Separation

Thin film composites of poly(2-hydroxyethyl methacrylate) and poly(methyl methacrylate) (PMMA) chain-tethered poly(vinylidene fluoride) (PVDF)-PDMS were prepared as a gas separation membrane (18). PDMS was coated on the PVDF support using a dip coating method. Poly(2-hydroxyethyl methacrylate) and PMMA were then grafted on PVDF-PDMS substrate by atom transfer radical polymerization.

The PVDF-PDMS-poly(2-hydroxyethyl methacrylate) and PVDF-PDMS-PMMA trilayer membranes were studied by attenuated total reflection FTIR spectroscopy, scanning electron microscopy (SEM), atomic force microscopy (AFM), water contact angle measurement, and X-ray photoelectron spectroscopy (XPS). The results of separation tests indicated that the CO_2/N_2 selectivity of PVDF-PDMS-poly(2-hydroxyethyl methacrylate) and PVDF-PDMS-PMMA thin film composites increased by ~2 and ~3 times, respectively, compared to the solvent-extracted PVDF-PDMS support (18).

1.2.4 *Adhesion Properties*

1.2.4.1 *Combinations with Thermoplastic Polymers*

The application field of combinations of thermoplastic polymers and LSR is remarkably wide, ranging from household consumers to the automotive sector (19). For the application of such combinations, the adhesion is a deciding factor.

In the case of LSR, tempering for improving its mechanical properties is used. However, there is no profound knowledge on annealing the combinations, thus making the influence on adhesion unclear.

The influence of different tempering temperatures (80°C, 100°C, and 120°C) and times (1, 3, 6, and 9 *h*) on the adhesion between the thermoplastic poly(butylene terephthalate) (PBT) and poly(amide) 12) and LSR was studied. The results showed that post-tempering influences the single components.

In the case of PBT, post-crystallization already occurs at 80°C, which is reflected, for instance, by increasing the degree of crystallinity by about 22%. LSR showed a post-crosslinking and the release of volatile components. Analyzing the tempering impact on adhesion, the peel resistance of LSR and PBT decreased around 23% at 80°C. The covalent bonds at the boundary layer were weakened due to the post-crystallization and the release of volatile components.

To identify the volatile components of the LSR, a gas chromatography coupled with mass spectrometry measurement was done. The test specimens were tempered in hermetic bags to catch the volatile components.

The volatile components mainly consisted of silicon-oxygen connections, such as siloxanes, silanes, and silanols, such as trimethylsilanol, disiloxane, cyclotritrisiloxane, and cyclotetrasiloxane, c.f. Figure 1.6.

During tempering, the volatile components are released. The amount of volatile components increased during varying tempering times, but their chemical composition remained the same while altering the temperature (19).

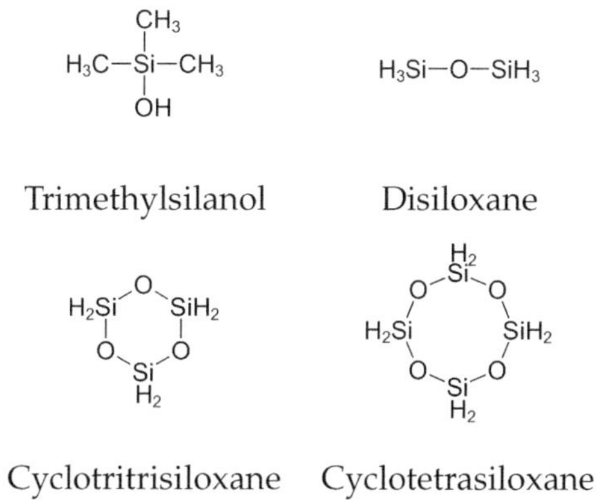

Figure 1.6 Volatile silicon-oxygen components (19).

1.2.4.2 *Octamaleamic Acid-POSS*

The addition of octamaleamic acid-polyhedral oligomeric silsesquioxane (POSS), c.f. Figure 1.7, nanoparticles to SR composites was investigated for enhancing the adhesion properties of reinforcing fibers (20). The content of octamaleamic acid-POSS was examined as the experimental parameter.

As reference a 5 *phr* silica-filled SR compound was used. The peroxide curing characteristics of composites was determined using differential scanning calorimetry (DSC) and a moving die rheometer.

It was found that octamaleamic acid-POSS retarded the curing time and decreased the crosslinking density. DSC analysis indicated that the crosslinking reaction started at a lower temperature as the octamaleamic acid-POSS concentration increased. The thermal mechanical analysis results showed that the melting point decreased with the addition of octamaleamic acid-POSS.

The crosslinking enthalpy and the curing onset temperature of SR composites are shown in Table 1.2.

The crystallinity of SR decreased with increasing octamaleamic acid-POSS content. The thermal stability of SR composites obtained from TGA apparently improved with the addition of octamaleamic acid-POSS. Scanning and transmission electron microscopy showed that the octamaleamic acid-POSS distribution was homogeneous at

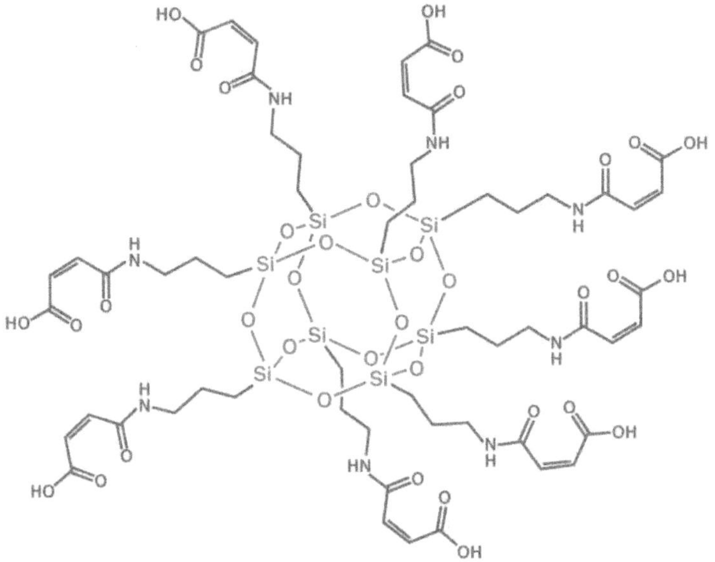

Figure 1.7 Octamaleamic acid-POSS.

Table 1.2 Crosslinking enthalpy and curing onset temperature of silicon rubber composites (20).

POSS added /[*phr*]	Curing enthalpy/[$J\,g^{-1}$]	Curing onset temperature/[°C]	Curing peak temperature/[°C]
0	13.0	150.7	189.3
1	7.4	151.4	181.2
3	8.5	151.8	183.4
7	7.6	145.3	187.8
10	8.2	145.5	188.2

lower contents, but some agglomerates were seen when the content of octamaleamic acid-POSS increased.

Tensile tests showed that comparable mechanical properties were achieved for SR/octamaleamic acid-POSS composites with respect to reference composite.

It could be demonstrated by H-adhesion tests that the utilization of octamaleamic acid-POSS in the SR composites improved the adhesion of the matrix to a Rayon fiber (20).

1.2.4.3 *Improvement of Adhesion Strength*

Polymeric hard/soft combinations consisting of a rigid, thermoplastic substrate and an elastomeric component offer many advantages for plastic parts in industry (21). Manufactured in one step by multi-component injection molding, the strength of the thermoplastics can be combined with sealing, damping or haptic properties of an elastomer. Bonds of self-adhesive LSR on high performance thermoplastics, such as poly(ether ether ketone) or poly(phenylene sulfide), are especially interesting for medical applications due to their outstanding resistance properties.

To ensure good adhesion between the two components, surface treatments from an atmospheric pressure plasma jet and a Pyrosil® flame were applied. The chemical changes on the thermoplastic surfaces were verified by water contact angle measurement and XPS.

A plasma treatment causes a decline in water contact angle, indicating the formation of functional groups, especially –OH, on the surface. XPS measurements confirm the increase of oxygen on the surface. Thus, the number of functional groups on the thermoplastic surface is enlarged by plasma treatment, leading to stronger bonding to the organofunctional silanes of the self-adhesive silicone rubber.

A thin layer of silanol groups is created by the Pyrosil flame on the thermoplastic substrates, which could be verified by XPS. A hydrophilic behavior of the coated surface is noticed. Both surface modification methods lead to enhanced adhesion properties of self--adhesive LSR on thermoplastic surfaces. This could be confirmed by 90° peel tests of the injection-molded composites, which led to an increase in the peel force by the surface modification techniques used (21).

1.2.4.4 *Tackiness Reduction*

A millable silicone rubber is widely used for producing rubber components because it can be mixed with a coloring agent with a simple device such as a twin-roll mixer and can be molded by simple equipment such as a press, and especially because its shape can be retained even in an uncured state. It is widely used in extrusion molding of tubes, gaskets, etc. (22).

However, the unique tackiness that the surface of a cured silicone rubber has is likely to cause the problem of blocking of surfaces of its molded products (22). Forming minute irregularities on the surfaces can prevent the blocking, and in molded products formed using a mold, the blocking can be easily prevented by roughening the surface of the mold. However, in extrusion molding and coating which do not use a mold, it is difficult to control the surface state.

A silicone rubber composition has been described that has improved moldability, in particular, extrusion moldability, while ensuring that a cured product obtained therefrom has sufficient strength. The silicone rubber composition contains (22):

A 100 part by mass base polymer consisting of a poly(organo siloxane diol) whose viscosity at 25°C is 1 to 100 $Pa\,s$ and a poly(organo siloxane) whose viscosity at 25°C is 0.2 to 40000 $Pa\,s$, with a ratio of the last compound of 20 to 100% by mass to the whole composition, the base polymer having a viscosity of 5 to 20000 $Pa\,s$ at 25°C and having an alkenyl group content of 0.001 to 0.3 $mmol\,g^{-1}$. Furthermore, a 10 to 50 part by mass silica powder whose specific surface area is 50 to 400 m^2g^{-1}, a 1 to 10 part by mass organosilazane, and a catalytic amount of a curing agent.

Several special examples for such compositions have been detailed (22). When such a composition is extruded at a rate of 2 $m\,min^{-1}$ by a screw extruder and thereafter cured into a tubular cured product, the tubular cured product preferably has, on a surface, 1 to 300 pieces/0.01 mm^2 granular protrusions whose maximum diameter measured in a scanning electron microscopic image is 0.1 μm to 30 μm. A SEM photograph (5000X) of the surface of the silicone rubber extrudate is shown in Figure 1.8.

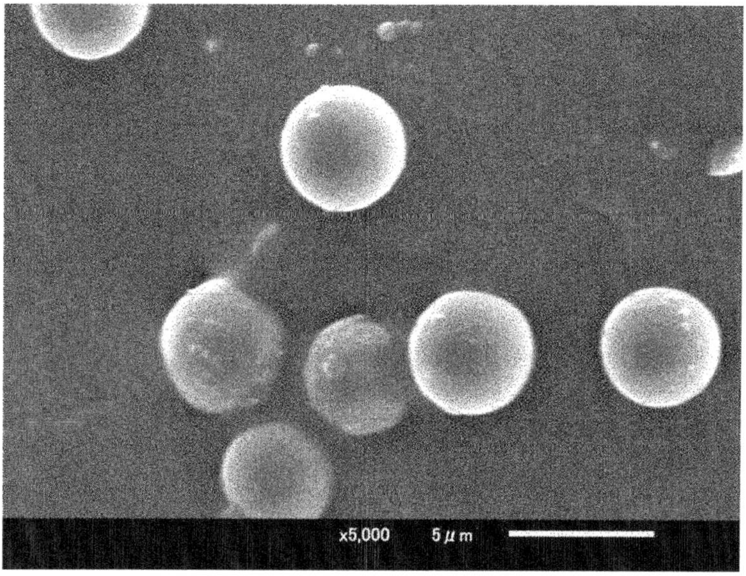

Figure 1.8 SEM photograph of a tubular cured product (22).

1.2.5 Pressure-Sensitive Adhesive Film

Silicone pressure-sensitive adhesives have the excellent heat resistance, freeze resistance and electrical properties inherent in silicone and maintain adhesion without impairing these properties (23). Thus, they can be widely used as pressure-sensitive adhesives where a high level of reliability is required.

Typical silicone pressure-sensitive adhesive compositions contain (23, 24):

1. A diorganopolysiloxane having at least two alkenyl groups in a molecule,
2. An organopolysiloxane containing $R_3SiO_{1/2}$ units and $SiO_{4/2}$ units in a molar ratio of the $R_3SiO_{1/2}$ unit to the $SiO_{4/2}$ unit of from 0.6 to 1.7, wherein R is a monovalent hydrocarbon group having 1 to 10 carbon atoms,
3. An organopolysiloxane having at least two silicon-bonded hydrogen atoms in a molecule,
4. An inhibitor,
5. A hydrosilylation catalyst, and
6. A solvent.

The silicone pressure-sensitive adhesive compositions may be applied to micro gravure coatings in order to obtain a thin pressure-sensitive adhesive layer on a substrate film. However, the aforementioned silicone pressure-sensitive adhesive compositions cannot form the pressure-sensitive adhesive layer exhibiting proper adhesion and good anti-scratch properties on the substrate film (23).

A silicone composition has been developed that exhibits proper viscosity and can form a pressure-sensitive adhesive layer that exhibits proper adhesion and good anti-scratch property on a substrate film (23). Another objective was to provide a pressure-sensitive adhesive film whose pressure-sensitive adhesive layer exhibits proper adhesion and a good anti-scratch property.

The compositions contain the ingredients shown in Table 1.3. These components will be explained in detail later on.

The alkenyl groups in component (A) can be exemplified by vinyl, allyl, butenyl, pentenyl, hexenyl, and heptenyl groups and are preferably vinyl groups. Component (A) can be exemplified by dimethylvinylsiloxy-endblocked dimethylpolysiloxanes, dimethylvinylsiloxy-endblocked dimethyl siloxane-methylvinyl siloxane copolymers, trimethylsiloxy-endblocked dimethyl siloxane-methylvinyl siloxane copolymers, partially branched chain dimethylpolysiloxane with molecular chain ends terminated by dimethylvinylsiloxy and trimethylsiloxy, trimethylsiloxy-endblocked partially branched chain dimethyl siloxane-methylvinyl siloxane copolymers.

Component (B) is another diorganopolysiloxane having at least one alkenyl group in a molecule. The alkenyl groups in component (B) can be exemplified by vinyl, allyl, butenyl, pentenyl, hexenyl, and heptenyl groups and are preferably vinyl groups. The non-alkenyl Si-bonded organic groups in component (B) can be exemplified by alkyl groups and aryl groups and are preferably methyl and phenyl groups.

The pressure-sensitive adhesive film is similar to that shown in Table 1.3.

The silicone composition exhibits a proper viscosity and can form a pressure-sensitive adhesive layer that exhibits proper adhesion and good anti-scratch property on a substrate film. Furthermore, the pressure-sensitive adhesive film has a pressure-sensitive adhesive

Table 1.3 Ingredients for pressure-sensitive adhesive compositions (23).

No.	Description
A	Diorganopolysiloxane having at least two alkenyl groups in a molecule, and having a viscosity at 25°C of from 10,000 to 1,000,000 *mPa s*, in an amount of from 60 to 80 mass based on a mass of the composition
B	A diorganopolysiloxane having at least one alkenyl group in a molecule, and being raw rubber-like at 25°C or having a viscosity at 25°C of more than 1,000,000 *mPa s*, in an amount of more than 0%, but not more than 10% based on a mass of the composition
C	An organopolysiloxane resin containing $(R^1_3SiO_{1/2})_x(SiO_{1/2})$, where R^1 is halogen-substituted or unsubstituted monovalent hydrocarbon group free from an alkenyl group and x is a number from 0.5 to 1.0, in an amount of from 0.5% to 20%
D	An organohydrogenpolysiloxane having at least two silicon-bonded hydrogen atoms in a molecule, in a quantity that provides from 0.1 to 10 moles of the silicon-bonded hydrogen atoms in this component per 1 mole of the total alkenyl groups in the composition
E	Silica fine powder in an amount of from 0.5 to 5% based on a mass of the composition
F	A hydrosilylation catalyst in a quantity that accelerates hydrosilylation of the composition

layer exhibiting proper adhesion and good anti-scratch property (23).

The pressure-sensitive adhesive film is produced by applying the silicone composition onto the surface of the substrate film, and then forming a pressure-sensitive adhesive layer on the surface of the substrate film by curing the composition at room temperature or with heating. Curing with heat is preferable. Heating should be carried out at a temperature above 50°C, preferably within the range of 80 to 200°C (23).

1.2.6 *Storage Stability*

A composition prepared by blending a resin microparticulate catalyst with a mixture of an organopolysiloxane and a crosslinking agent cannot secure long-term storage stability in a mixed state at around room temperature depending on the type of the resin making up the resin microparticulate catalyst, so that problems of a significant increase in viscosity of the mixture and progress of curing of the mixture are caused (25).

In addition, it takes time to crosslink the mixture when curing the mixture by heating, so that the composition could show a low crosslinking reactivity. Until now, there has not been a technique for obtaining a silicone rubber composition that has both excellent storage stability in a mixed state and excellent crosslinking reactivity during heating.

A silicone rubber composition has been developed that contains an organopolysiloxane, a crosslinking agent, and a microcapsule type of catalyst that is made of microparticles of a resin and a crosslinking catalyst encapsulated in the microparticles.

Examples of the organopolysiloxane include an alkenyl group-containing organopolysiloxane, a hydroxyl group-containing organopolysiloxane, a (meth)acryl group-containing organopolysiloxane, an isocyanate-containing organopolysiloxane, an amino group-containing organopolysiloxane, and an epoxy group-containing organopolysiloxane.

The alkenyl group-containing organopolysiloxane is used as a main material for an addition curing type silicone rubber composition. The alkenyl group-containing organopolysiloxane is crosslinked by a hydrosilyl crosslinking agent in addition reaction with the hydrosilyl crosslinking agent. While proceeding even at room temperature, this addition reaction is promoted under heating. Thermal curing by this addition reaction is preferably performed at at 100–170°C.

The preparation of microcapsule type catalysts can be done as follows (25):

Preparation 1–1: A toluene solution of a platinum catalyst, containing 3% by mass of platinum metallic atoms, a coating resin for microparticulation, and toluene were mixed at a mass ratio of 0.6:5:95, and the thus-prepared solution was dropped into a water solution of a surface acting agent to prepare an emulsion. Then, the toluene was distilled and removed under reduced pressure and the emulsion was filtered to obtain microparticles for each catalyst that contain the coating resin and the platinum catalyst. As platinum catalyst, platinum chloride (IV) manufactured by Furuya Metal Co., Ltd. was used.

1.2.7 *Thermal Stability*

A platinum catalyst and a nitrogen-containing silane were introduced into a silicone rubber to improve the thermal stability (26). The effects of Pt and nitrogen-containing silane on the thermal stability and degradation mechanism of silicone rubber were investigated by thermogravimetry (TG), TG-FTIR spectrometry, SEM and other methods.

A significant synergism was found between Pt and nitrogen-containing silane for improving the thermal stability of the silicone rubber. When 6.67 *ppm* of Pt and 1.4 *phr* of nitrogen-containing silane were introduced, the temperature of 10% and 20% weight loss under nitrogen atmosphere were respectively increased by 36°C and 119°C. The residue weight at 900°C was doubled to 68% in the presence of the Pt/nitrogen-containing silane.

The synergistic mechanism might be that the nitrogen atom coordinated with Pt and improved the catalytic efficiency of Pt. Additionally, nitrogen-containing silane preserved the catalytic activity of Pt under air atmosphere. Thus, the Pt/nitrogen-containing silane efficiently catalyzed the thermal crosslinking and suppressed the degradation of silicone chains. Moreover, it revealed that the presence of Pt/nitrogen-containing silane protected silicone chains from oxidation. Thus, the unzipping depolymerization by silanol groups was reduced significantly (26).

1.2.8 *Hydrophobed Pyrogenic Silica Filler*

In order to prevent the undesirable influence of the silanol groups on the mechanical properties of a silicone rubber, it is necessary to render the surface of the pyrogenic silica hydrophobic (27).

A silicone rubber composition has been described, which is characterized by its containing a structurally modified hydrophobic pyrogenic silica as filler (28).

The silanized, structurally modified silica has vinyl groups fixed to the surface, hydrophobic groups such as trimethyl silyl or dimethyl silyl and/or monomethyl silyl groups additionally being fixed to the surface. The preferred properties are shown in Table 1.4.

Table 1.4 Preferred properties of the composition (28).

Property	Amount
BET surface area	25 – 400 $m^2 g^{-1}$
Average primary particle size	5 – 50 nm
pH	3 – 10
Carbon content	0.1 – 10 %
DBP value	<200 %

The dibutyl phtalate (DBP) value in Table 1.4 is defined as the volume of dibutyl phtalate absorbed by 100 g of black pigment (29,30).

1.2.9 *Superhydrophobic Materials*

1.2.9.1 *Superhydrophobic Surfaces*

The fabrication of large-scale superhydrophobic surfaces for commercial applications is challenging due to certain limitations. A simple and inexpensive method has been developed to fabricate superhydrophobic surfaces on silicone rubbers (31).

Templates with different rough structures were prepared first. Silicon carbide particles of (63, 21, 15, and 10.5) μm in diameter were obtained by using test sieves of 80, 240, 600, 800, and 1200 *mesh*, respectively. A uniform layer of resin binder-epoxy resin was sprayed on the inner surface of a cubic mold with dimensions of $3cm \times 3cm \times 3cm$ after treating with a silane coupling agent

N-β-aminoethyl-γ-aminopropyl trimethoxysilane. The prepared silicon carbide particles were evenly sprayed on the inner surface of the mold and dried at room temperature. Then rough silicon rubber samples were prepared through a conventional molding process.Liquid silicone rubber was poured into the mold, the mold was taken off after consolidation, and thus silicone-rubber surfaces with different surface morphologies were obtained (31).

Rough microstructures were prepared on the inner surfaces of the molds and then sample superhydrophobic surfaces on silicone rubbers with different surface roughness were achieved using the standard molding process.

Furthermore, the effects of roughness on the wettability were investigated. The results showed that by controlling the roughness, the fabricated surfaces exhibited a static contact angle of 150.9° and a sliding angle of 8°. Finally, the property of hydrophobicity recovery for the silicone-rubber samples was studied. The surfaces of the samples could recover well after a sandblasting experiment. The proposed method is low cost, environmentally friendly and suggests promising industrial applications (31).

1.2.9.2 *Flashover Characteristics*

The influence of the low adhesion of superhydrophobic surface on flashover characteristics under wet conditions has been studied (32).

The samples are prepared as follows (32,33):

Preparation 1–2: A sheet of glass with dimensions of 25 *mm* ± 75 *mm* ± 1 *mm* was put into a mold. The mold was 25 *mm* ± 75 *mm* ± 2 *mm*. Then, a liquid one-component room temperature vulcanized silicone rubber was poured into the mold. After the liquid silicone rubber cured, the common silicone rubber sample was acquired. The thickness of room temperature vulcanized coating was about 1 *mm*.

The common silicone rubber sample was similar to glass insulators coated by room temperature vulcanized coating. Then a mixture of ethyl acetate, liquid silicone rubber and SiO_2 nanoparticles was sprayed on the common silicone rubber sample. After the silicone rubber cured, a superhydrophobic surface was formed on the silicone rubber sample. Here, a superhydrophobic silicone rubber sample was acquired. The contact angle of superhydrophobic surface was larger than 150° and the sliding angle was less than 2°.

The flashover experiments were done with copper foils with 0.05 *mm* thickness that were used as electrodes. The copper foils were glued onto the samples. The distance between the electrodes was 2.5 *cm*. Then the electrodes were connected to an AC high voltage source. The mode of increase of voltage was a step-by-step test. If the samples withstood a voltage for 2 *s* without failure, the voltage would increase in steps of 0.2 *kV*. The voltage would be increased until flashover occurred. The whole process was recorded by a digital camera.

The flashover experiments were conducted under two different wet conditions. One was placing a constant volume droplet on the silicone rubber. The other one was that silicone rubber was wetted by salt fog.

It was found that the adhesion between water droplets and a superhydrophobic surface was very low, because of the presence of air cushion. Accordingly, water droplets were easy to slide on a superhydrophobic surface under the effect of electric filed. The sliding of droplets could provide a longer insulation path before the flashover occurred. The results of the study showed that the flashover voltage could be improved greatly on a superhydrophobic silicone rubber surface (32).

1.2.10 Thermally Conductive Materials

1.2.10.1 Layer-by-Layer Assembly

Thermally conductive silicone rubber composites used for heat removal from electronic devices have attracted great attention (34). A facile spin-assisted layer-by-layer assembly approach was used to fabricate highly thermally conductive multilayered silicone rubber/graphene films.

The films exhibit a highly ordered lamellar structure with the high orientation of graphene. This provides continuous thermally conductive pathways in the horizontal direction (34).

A multilayered film with 40 assembly cycles has the thermal conductivity of 2.03 $W(mK)^{-1}$ in in-plane direction. Moreover, the film can be highly twisted to any angle and has an elongation at break of 325%, which is rarely achieved in previously reported graphene-based multilayered films.

Even up to 500 stretch-recovery cycles at 50% strain, the change of the thermal conductivity was negligible, indicating the high durability, excellent flexibility and stretchability of the film. The films with high thermal conductivity and stretchability have potential applications in flexible electronics, wearable devices and electronic skin (34).

1.2.10.2 Boron Nitride

Boron Nitride Nanosheets. An efficient method was reported to fabricate boron nitride nanosheets using a sonication technique in mixed solvents (35). The functionalization of boron nitride nanosheets is performed by poly(dopamine). The boron nitride nanosheets showed good dispersibility in mixture solvent. And the functionalization of boron nitride nanosheets by poly(dopamine) can further improve their stability in water.

The morphologies of the boron nitride nanosheets and the poly(dopamine) modified compositions were characterized by AFM, SEM, and transmission electron microscopy, which confirmed the successful exfoliation and functionalization of boron nitride nanosheets. Furthermore, silicone composites were fabricated by adding boron nitride nanosheets and boron nitride nanosheets/poly(dopamine), respectively.

Compared with neat silicone (0.20 $W m^{-1} K^{-1}$), the thermal conductivity of silicone composite with 12% BNNSs poly(dopamine) increases highly by 90% to 0.38 $W m^{-1} K^{-1}$. The thermal stability of silicone/boron nitride nanosheets/poly(dopamine) composite is also improved.

The significantly improved thermal conductivity and thermal stability can be attributed to the good dispersion of boron nitride nanosheets/poly(dopamine) and a stronger interaction between the boron nitride nanosheets/poly(dopamine) surface and the silicone matrix (35).

Vertically Aligned Boron Nitride. Silicone rubber/vertically aligned boron nitride composites were prepared and the thermal conductivity of the composites was systematically investigated (36). The vertically aligned boron nitride molecules constructed effective vertical thermal conductive pathways in the SR matrix, and so the composites exhibited much higher through-plane thermal conductivity compared to the non-oriented composite.

The through-plane thermal conductivity of the composites reached 5.4 $kW\,m^{-1}K^{-1}$, which was 33 times higher than those of pure SR. In addition, the surface thermal infrared analysis also indicates that the vertically aligned SR/boron nitride composites possess an excellent heat transfer capacity during the heating and cooling processes (36).

1.2.11 *Shape-Memory Materials*

Shape-memory polymers are a kind of stimuli-responsive materials that can have physical transformation between their initial shape and temporary shape under certain external stimulus, such as heat, electricity, light chemical induction, and others. Shape-memory polymers have several advantages (37–39).

Many kinds of stimulation can induce the shape-memory effect. Furthermore, they can be designed in a broad range of ways, and have a flexible application (40).

A shape-memory polymer (SMP) contains two structure elements (41): A permanent network and reversible network. The permanent network is an elastic network, which is the part that keeps the permanent shape and drives recovery upon the application of a stimulus. The reversible network is able to counterbalance the load stored in the permanent network to fix a temporary shape and is sensitive to the stimulus-allowing recovery when the network is reversed.

The thermally induced SMP is mainly derived from the reversible network that undergoes a thermal transition such as a melting or glass transition (42, 43).

An electro-active SMP is a composite of thermally induced SMP materials and a conductive material, such as graphitized carbon black, or a metallic powder. The mechanism is similar to that of the thermally induced SMP. The conductive material increases the temperature of the system through the heat generated by the current, resulting in the shape recovery, so it has both conductive and shape-memory effects.

Four fillers were investigated in a study (40): Cornstarch,

1,10-decanediol, 12-hydroxystearic acid, and stearic acid. These compounds are shown in Figure 1.9.

By adding cornstarch, the sticky silicone is somewhat stiffened and starts to quickly set up from the inside out because of its moisture content. The other three fillers all can produce a shape-memory effect when they are blended with silicone rubber (40).

However, the performance is different. Compared with other samples, the 12-hydroxystearic acid blend sample does not perform well, its recovery is only around 80%. The decanediol silicon rubber blend showed the best performance; the next composition was stearic acid, with 12-hydroxystearic acid showing the poorest performance.

By mixing silicone rubber with 1,10-decanediol the shape-memory effect can be realized. The crystalline small 1,10-decanediol molecule could provide the reversible network. The silicone rubber part results in an elasticity. Samples can be stretched to 200% elongation and display good shape-memory properties. According to the shape-memory cycle test, a 50% 1,10-decanediol silicone rubber composition shows the best performance in terms of high fixity and recovery, and it displays good repeatability after several shape-memory cycles. By comparing 1,10-decanediol with other crystalline small molecules, 1,10-decanediol silicone rubber composition displayed the best performance (40).

1.2.12 Thermally Conductive Grease

Electronic parts generate heat in use and, thus, the performance thereof lowers owing to the heat generation. In order to solve this problem, a variety of heat dissipation techniques have been employed. One of the ordinary methods includes placing a cooling member in the vicinity of a heat generation unit and bringing them in intimate contact with each other, thereby dissipating heat by efficient removal of heat through the cooling member.

In this connection, however, if there is a space between the heat generation member and the cooling member, thermal conduction does not proceed smoothly because of the presence of air, which has poor thermal conductivity, so that the temperature of the heat generation member does not decrease satisfactorily. To prevent this, for the purpose of preventing the presence of air, heat dissipating

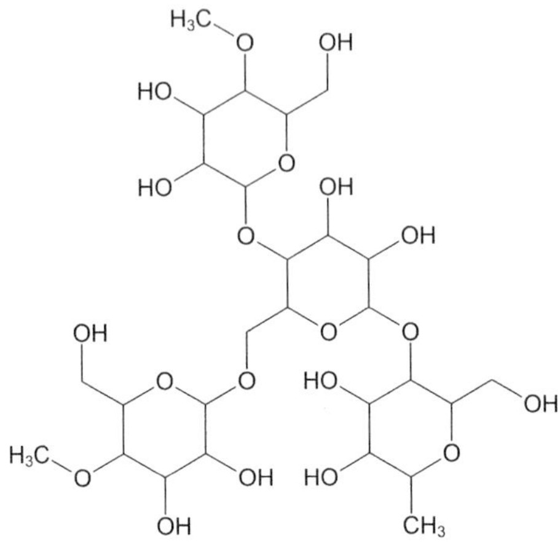

Cornstarch

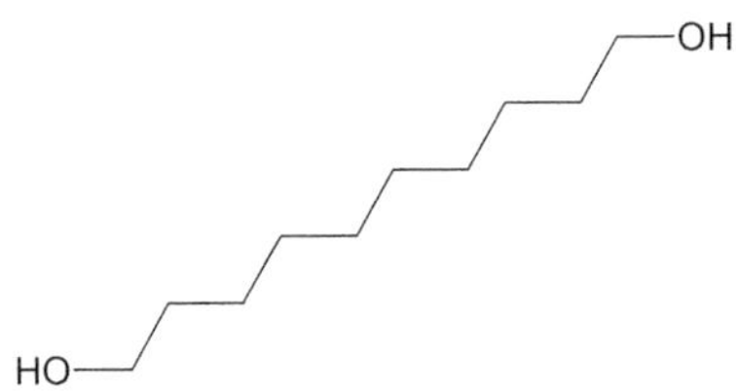

1,10-Decanediol

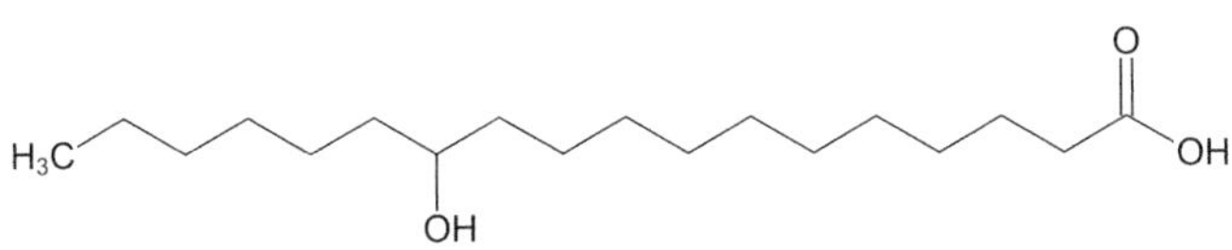

12-Hydroxystearic acid

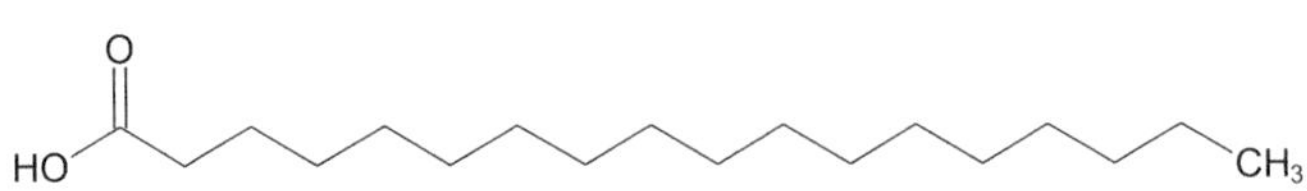

Stearic acid

Figure 1.9 Fillers.

materials, heat-dissipating sheets or heat dissipating greases have been conventionally used, which are good at thermal conductivity and are able to follow the surface profile of the member.

A thermally conductive grease composition has been developed, which scarcely increases in hardness during high-temperature heating and has a minimized reduction in growth (44). The components of the composition are:

1. An alkenyl group-containing organopolysiloxane, examples are dimethylpolysiloxanes blocked with a dimethylvinylsilyl group at both ends.
2. A hydrolyzable methylpolysiloxane, trifunctionally terminated at one end,
3. A thermally conductive filler,
4. An organohydrogenpolysiloxane,
5. A triazine ring and alkenyl group-containing bonding aid, and
6. A platinum catalyst, and, if required,
7. An addition reaction inhibiting agent.

An example of a hydrolyzable methylpolysiloxane compound is shown in Figure 1.10. Examples of the organohydrogenpolysiloxane compounds are shown in Figure 1.11. A triazine compound is shown in Figure 1.12

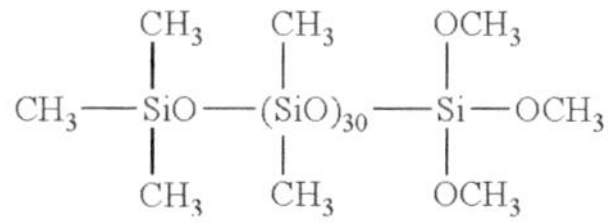

Figure 1.10 Hydrolyzable methylpolysiloxane (44).

1.2.13 Self-Healing Materials

Graphite nanoplatelets recently attracted attention as a viable and inexpensive material that can be used in many engineering applications, given the excellent in-plane mechanical, structural, thermal, and electrical properties of graphite (45–47).

Composites with carbon-based fillers show thermal stability, light weight, and a high thermal conductivity (48,49).

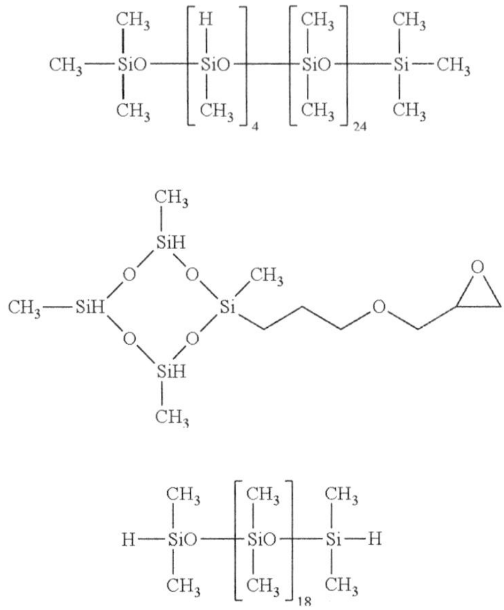

Figure 1.11 Organohydrogenpolysiloxanes (44).

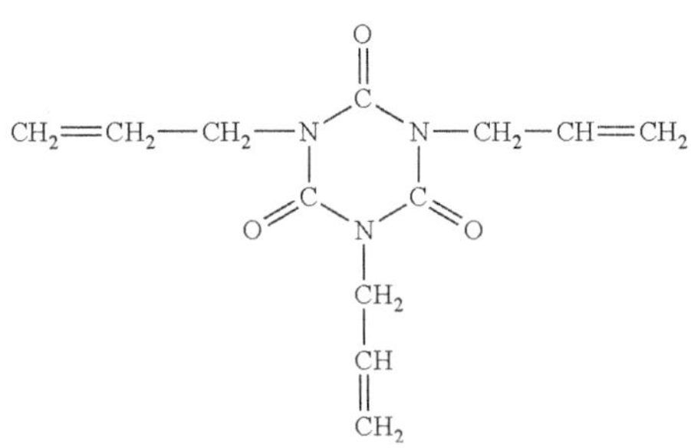

Figure 1.12 Triazine compound (44).

These properties suggested that integrating graphite nanoplatelets into SR might generate novel self-healing materials, which have multifunctional properties such as self-healing by thermal annealing and temperature activation of the electrical conductivity (50).

1.2.14 *Flame Retardancy*

To effectively enhance the flame retardancy of silicone rubber is a challenging task. Functionalized zirconium phosphate was prepared by the intercalation of 1,2,2,6,6-pentamethyl-4-(vinyldiethoxysiloxy) piperidine into a zirconium phosphate, c.f. Figure 1.13, and was employed to enhance the flame retardancy of addition-cured LSR (51).

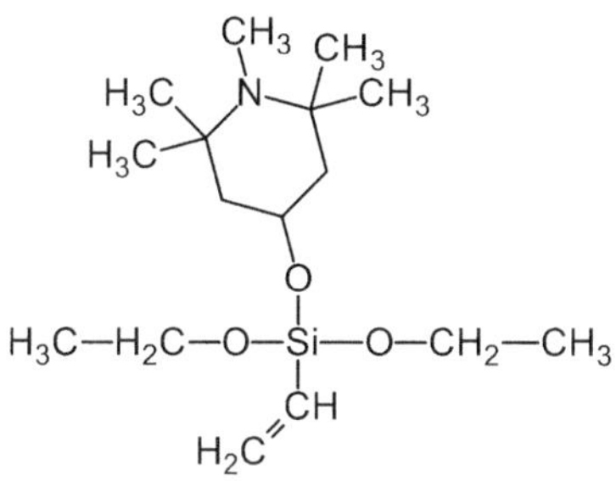

Figure 1.13 1,2,2,6,6-Pentamethyl-4-(vinyldiethoxysiloxy) piperidine.

The results of the study showed that addition-cured LSR had no rating in the UL-94 test (52,52) with the limiting oxygen index (LOI) value of 28.0%.

Functionalized zirconium phosphate could significantly improve the flame retardancy of addition-cured LSR. When only 4.0 *phr* functionalized zirconium phosphate was added, the LOI value reached 31.0% and the vertical burning test (UL-94) passed the V-0 rating. Moreover, compared to addition-cured LSR, the peak heat release rate and the total heat release of addition-cured LSR/functionalized zirconium phosphate decreased by 54.2% and 47.5%, respectively. The probable flame retardant mechanism of functionalized zirconium phosphate was mainly attributed to the synergism of the catalyzing carbonization and free-radical quenching effects. On the one hand, 1,2,2,6,6-pentamethyl-4-(vinyldiethoxysiloxy) piperidine suppressed the thermal degradation of silicone chains in the condensed phase and quenched the active free radicals in the gas phase by generating nitroxyl radicals.

On the other hand, ZrP promoted silicone chain self-crosslinking and the formation of the graphitized carbons, resulting in the formation of the compact and thermostable ceramic layer, which was a good barrier to protect the underlying matrix and inhibit the exchange of degradation products and oxygen (51).

In another study, the effect of POSS particles and the synergism of POSS and poly(melamine phosphate) on the thermal properties and flame retardancy of silicone rubber composites was investigated (53).

Flame retardant compounds for SR in the form of polyhedral POSS, containing both isobutyl groups and amino-propyl groups, or chloro-propyl groups or vinyl groups were tested. The structures are shown in Figure 1.14.

The silsequioxanes were incorporated into the silicone rubber matrix in a quantity of 3 and 6 parts by weight by the method of reactive stirring with the use of a laboratory mixing mill.

Based on TGA, FTIR and various SEM methods, the thermal degradation mechanism of non-crosslinked and crosslinked silicone rubber was elucidated. The effects of POSS, and POSS in a synergic system with melamine poly(phosphate), c.f. Figure 1.15, on the thermal properties and flammability of silicone rubber composites were reported.

Also, a mechanism of the flame retardant action has been proposed. The thermal degradation of silicone rubbers can proceed according to three reaction mechanisms: Unzipping, random scission of the Si–O bond in the main chain of polymer, and an external catalysis. PDMS containing silanol (Si–OH) or hydroxyl-alkyl-silanol (Si–R–OH) terminal groups, undergo thermal degradation by an unzipping mechanism. In the first stage of heating, the molecular weight of the polymer increases as a result of an intermolecular condensation process between the terminal and the silanol groups and interchain Si–O groups. A further increase in temperature causes a linear decrease in the polymer molecular weight as a result of the formation of cyclic siloxanes followed by their tearing off and passing to the gas phase of cyclic siloxanes, mainly trimers and tetramers.

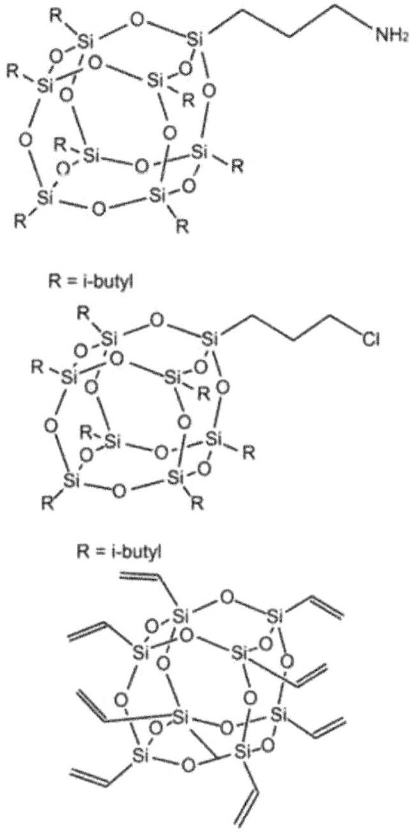

Figure 1.14 Polyhedral POSS compounds (53).

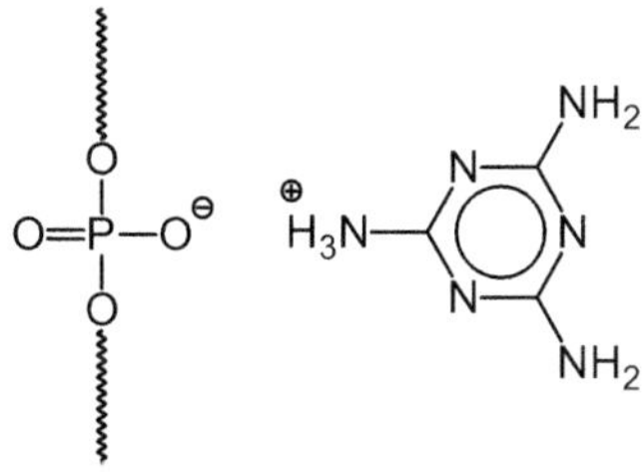

Figure 1.15 Melamine poly(phosphate).

It has been shown that POSS, especially with melamine poly(phosphate), considerably increases the thermal stability and decreases the flammability of the SR rubber composites (53).

1.3 Special Materials

1.3.1 *Borosilicones and Viscoelastic Silicone Rubbers*

Borosilicones, e.g., Silly Putty, have been known for 70 years, but their peculiar behaviors have remain unexplained. Experiments and theory were used to show that they are network liquids—dynamic macromolecules that appear elastic on short timescales but exhibit flow on longer timescales (54).

There are a number of techniques for incorporating boron in poly(organo siloxane)s (55) and consequently for producing silicone bouncing putties (56,57).

The reaction between a silanol group and a methoxy group on boron is a condensation reaction (54):

$$\begin{array}{c} B(O{-}CH_3)_3 + HO{-}PMDS{-}OH \rightarrow \\ B(O{-}CH_3)_2{-}O{-}PMDS{-}OH + CH_3OH \end{array} \qquad (1.1)$$

This reaction is reversible, so that alcoholysis accompanies condensation, and it normally reaches an equilibrium in which both reactants and products are present.

To drive the reaction toward condensation and borosilicone formation, the methanol must be removed. Simply allowing the methanol to evaporate is sufficient, but vacuum drying propels the reaction to completion far more efficiently. The condensation reaction is so rapid, even at room temperature, that vacuum removal of the methanol can cause the borosilicone to form within seconds (54).

Boron-containing crosslinking agents may be boric acid or a boric acid ester such as trimethyl borate, triethyl borate, triisopropyl borate, and tributyl borate (57). Due to their chemical structure, boron-containing crosslinking agents have three functional groups which can react with the silanol groups in the poly(organo siloxane). The use of a boron compound as the temporary crosslinking agent has an additional consequence: This material typically exhibits a remarkably high stiffness and resiliency in response to sudden impacts.

Chemicals bearing hydroxyl and carboxyl groups are particularly effective boron bridge-opening chemicals. Water, alcohols, polyols, silanols, and carboxylic acids are examples of chemicals bearing hydroxyl and carboxyl groups. Some chemicals are collected in Table 1.5 and some chemicals are shown in Figure 1.16.

Table 1.5 Boron bridge-opening chemicals (57).

Primary alcohols	Carboxylic acids
2-Propanol	Acetic acid
Hexanol	2-Ethylhexanoic acid
Decanol	Lauric acid
2-Ethylhexanol	Stearic acid
Lauryl alcohol	Oleic acid
Stearyl alcohol	Isostearic acid
Oleyl alcohol	
Isostearyl alcohol	

Here, simple borosilicones were prepared by reacting a silanol-terminated poly(dimethyl siloxane) fluid HO-PDMS-OH and related compounds with the liquid boron compound trimethyl borate $B(OCH_3)_3$ (54). Examples of such poly(siloxane)s are shown in Figure 1.17.

Each borosilicone is a vast covalent network of silicone polymer chains joined by trifunctional boron crosslinks. At any instant, the borosilicone is a highly crosslinked elastic material. Because the boron crosslinks are temporary, however, the network evolves with time and the borosilicone exhibits a liquid behavior.

A simple borosilicone exemplifies a classic transient network model and behaves as a simple (Lodge) elastic fluid. Its measured moduli and viscosities fit those predicted by the transient network model and the Maxwell viscoelastic model: A spring in series with a dashpot, including the observed exponential relaxation processes.

When a borosilicone includes permanent crosslinks, however, it no longer behaves as a simple elastic fluid. Its measured moduli and viscosities fit those predicted by the fractional Maxwell viscoelastic model: A spring in series with a spring-pot, including the observed slower-than-exponential relaxation processes.

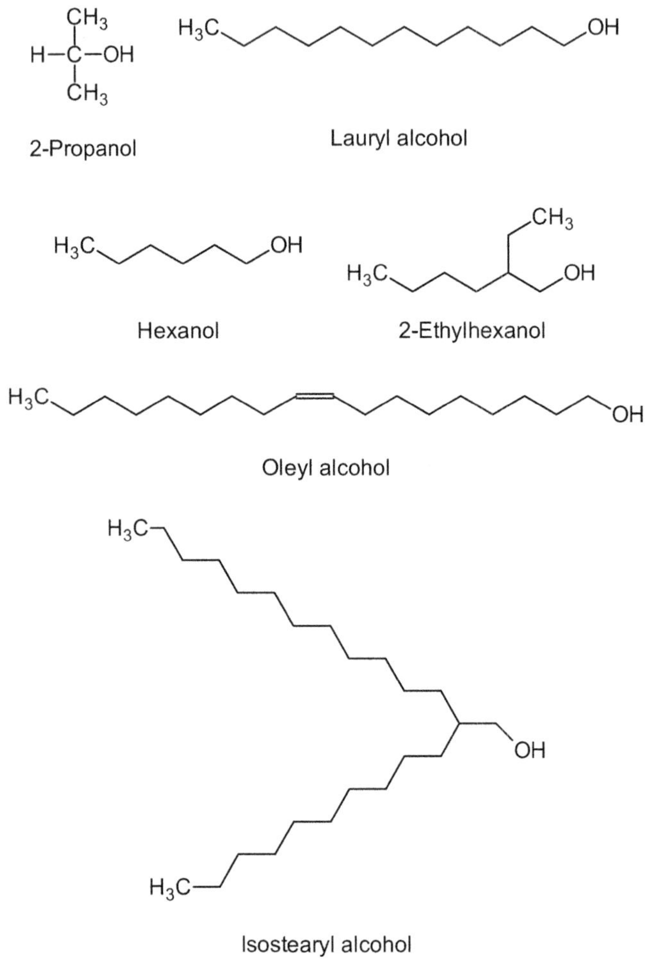

Figure 1.16 Boron bridge-opening chemicals.

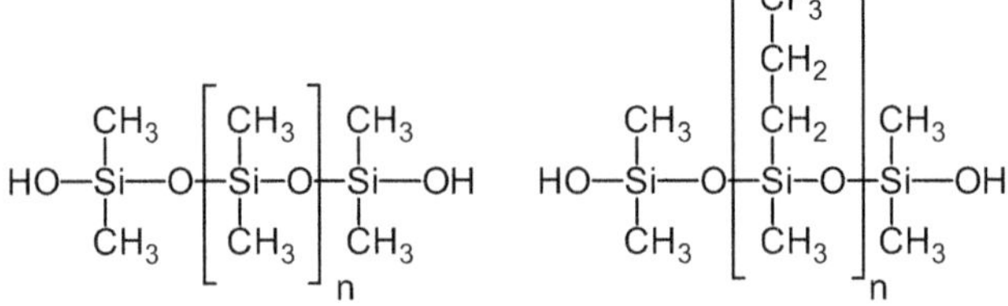

Figure 1.17 Silanol-terminated poly(siloxane)s.

Beyond the gelation threshold, a borosilicone becomes a viscoelastic silicone rubber (VSR). With a permanent network that spans the material coupled to a temporary network that also spans the material, the VSR is a network liquid piggybacking on a network solid. The fractional Zener viscoelastic model: An elastic spring in parallel with the fractional Maxwell model accurately predicts the measured moduli of VSRs. The temporary nature of boron crosslinks is due to exchange reactions. Because the mean lifetime of temporary crosslinks is a borosilicone's only significant timescale, it exhibits thermorheological simplicity (54).

1.3.1.1 *One-Pot Synthesis*

A one-pot synthesis procedure of polyborosiloxane resins by dehydrocarbon condensation of trimethyl borate with diphenylsilane in the presence of tris(pentafluorophenyl)borane as catalyst has been reported (58). The condensation process, which leads to the formation of Si–O–B linkages with a release of methane, was performed in a hydrocarbon solvent under mild conditions. The proposed reaction scheme is shown in Figure 1.18.

The so fabricated polyborosiloxane glassy product is halogen-free and has a low content of hydroxyl groups. High temperature decomposition of the polyborosiloxane resin yields a silicon-boron-oxycarbide ceramic with a high yield above 54%, when the process is carried out in a nitrogen atmosphere. The char yield is about 10% lower for thermal decomposition in the presence of air rather than in nitrogen.

The polyborosiloxane resins obtained by this process could find applications as high temperature adhesives, precursors to ceramic materials and halogen-free flame retardants (58).

1.3.1.2 *Interfacial Adhesion*

An attempt was made to improve the interfacial adhesion property between epoxy resin (EP) and addition-curable liquid silicone rubber (59).

As a surface treatment agent, borosiloxane oligomers were synthesized, and the major chemical structure was evaluated by gel per-

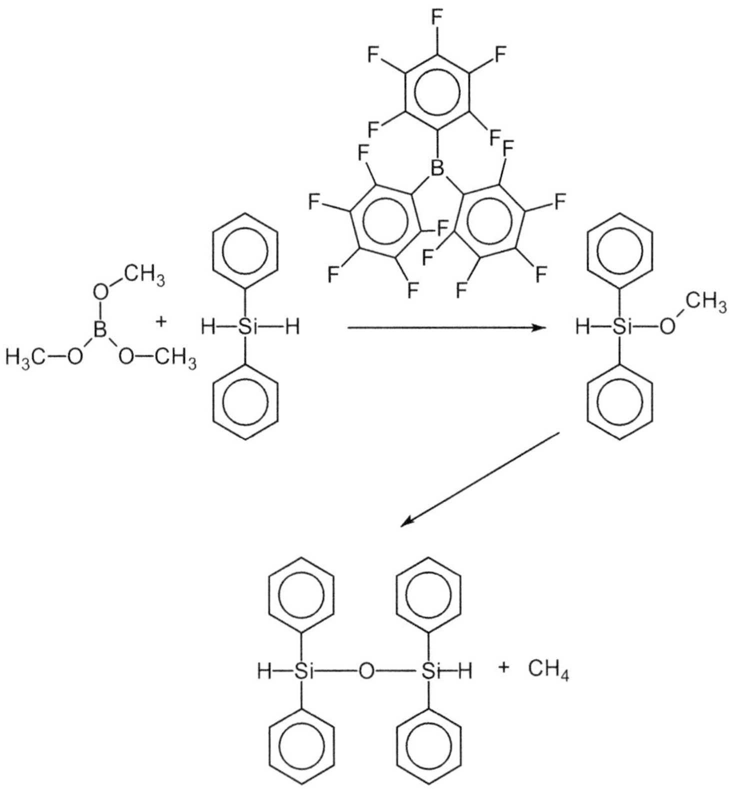

Figure 1.18 Formation of Si-O-Si bridges via exchange of functional groups and subsequent dehydrocarbon condensation (58).

meation chromatography, FTIR, nuclear magnetic resonance spectroscopy (NMR), and TGA.

Borosiloxane oligomers served as surface treatment agent for an epoxy resin to improve the interfacial adhesion characteristics of the epoxy resin and addition-curable liquid silicone rubber (59). The adhesion strength between epoxy resin and addition-curable liquid silicone rubber was measured by a pull-out test. The results indicated that the adhesion strength increased from 0.02 *MPa* for an epoxy resin/addition-curable liquid silicone rubber to 0.18 *MPa* for a borosiloxane oligomer-epoxy resin/addition-curable liquid silicone rubber, with the failure mode from epoxy resin posts dislodgement for epoxy resin/addition-curable liquid silicone rubber to rubber breakage for borosiloxane oligomers-epoxy resin/addition-curable liquid silicone rubber.

Furthermore, the adhesive mechanism was analyzed and it was concluded that the improvement of adhesion characteristics of the epoxy resin/addition-curable liquid silicone rubber through the oligomer surface treatment was a result of chemical and physical bonding. The borosiloxane oligomers acted as binder by reacting with amino groups on the surface of the epoxy resin through the active epoxy groups and absorbing electron clouds of the phenyl groups of the addition-curable liquid silicone rubber chains through the boron atoms.

This feature makes borosiloxane oligomers a superior candidate to improve the adhesion and reduce the dislodgement or the exfoliation of the silicone rubber or other materials possessing electron-rich groups on the molecular chains from epoxy resin (59).

1.3.2 *Acrylo-Polyhedral Oligomeric Silsesquioxane*

Acrylo POSS, a liquid POSS derivative with a reactive C=C double bond, was used to modify addition-cured LSR as an effective nanofiller (60).

Significant enhancements on mechanical properties were obtained. With the addition of only 1.5 *phr* of acrylo POSS to fumed silica-strengthened silicones, the Young's modulus and ultimate tensile strength are increased by 432% and 66%, respectively, and the hardness of resulting LSR composites was improved. NMR and FTIR spectroscopy proved an efficient hydrosilylation between acrylo POSS and hydrosiloxane directly.

The thermal stability and morphology investigations also confirmed that POSS is covalently incorporated into the network of the silicone rubber. The increment of crosslink density was proved by extraction and swelling experiments and dynamic mechanical analysis. It could be envisioned that this simple and effective method could help to produce high performance silicone rubber composites (60).

1.3.3 *Cellulose Nanocomposites*

To prepare LSR/cellulose nanocrystal nanocomposite, 3-aminopropyltriethoxysilane was used to modify the surface of the cellulose nanocrystal to improve the interfacial interaction between the hydrophilic cellulose nanocrystal and the hydrophobic LSR (61).

FTIR spectroscopy and energy dispersive spectrometry results demonstrated that 3-aminopropyltriethoxysilane modified the cellulose nanocrystal successfully. It was found that a small amount of modified cellulose nanocrystal had better reinforcement than SiO_2 in LSR nanocomposites; the tensile strength and the strain at break of LSR increased over 85% and 44% by adding 1.5% modified cellulose nanocrystal.

However, DSC and rheology tests indicated that 3-aminopropyltriethoxysilane modified cellulose nanocrystals increased the curing temperature of LSR and limited the addition of a cellulose nanocrystal, which was attributed to poisoning of the platinum catalyst by 3-aminopropyltriethoxysilane. Furthermore, a small amount of modified cellulose nanocrystal improved the thermal stability of LSR; the TGA results showed that the 10% weight loss temperature of LSR increased 56°C with a 1.0% addition of modified cellulose nanocrystals (61).

1.3.4 *Fluorine-Containing Poly(phenylsilsesquioxane)*

It is of great interest to simultaneously improve the anti-pollution and tracking resistance of a silicone rubber for its application in outdoor high voltage insulation (62).

An efficient approach was proposed to address this issue by incorporating fluorine-containing poly(phenylsilsesquioxane) and an ureido-containing MQ (M= $(CH_3)_3SiO$, Q= SiO_4) silicone resin (U-MQ) into addition-cure LSR. The fluorine-containing poly(phenylsilsesquioxane)/U-MQ composition significantly improved the tracking resistance and hydrophobicity of the addition-cure LSR. By adding 2 *phr* fluorine-containing poly(phenylsilsesquioxane) and 1 *phr* U-MQ, addition-cure LSR passed the 1A 4.5 level inclined plane test with merely 0.2% eroded mass, and had a water contact angle as high as 121 Å. The synergism of fluorine-containing poly(phenylsilsesquioxane) and U-MQ on suppressing the addition-cure LSR degradation was further revealed and demonstrated. These findings exhibited great potential for fabricating silicone rubber with prominent tracking and erosion resistance by the combination of prevention and inhibition of tracking damage, endowing it with excellent long-term stable service performance (62).

1.3.5 Silicone Rubber Overmolded Poly(carbonate)s

Poly(carbonate)s (PCs) and blends of PC and polyester are utilized in a wide variety of molded structures, for instance, automotive parts, medical devices, medical device components, electronic device components such as cell phones, glasses, and goggles, such as virtual reality googles and safety goggles.

These materials are forming structures that have rigid surfaces. Rigid surfaces can be irritating to users. To reduce or prevent structures formed from PCs from irritating users, softer materials such as cured silicon rubbers are overmolded onto the surfaces or a portion of the surfaces of such structures.

However, cured silicon rubbers do not bond well to the surfaces of molded PCs (63). A number of references disclose methods of enhancing the bonding of silicon rubber in the cured state to PCs (64–66).

The market demands cost-effective solutions that further enhance the bond strength of silicon rubber to molded PCs (63).

Articles have been published which include the following solutions (63): A molded substrate consisting of one or more PC resins or blends of PC and polyester resins containing poly(dimethyl siloxane)s modified with acrylate groups, hydroxyl groups, or epoxy groups. These are disposed on the surface. Or a portion of the substrate is a cured silicone rubber; PCs contain the modified poly(dimethyl siloxane)s in a sufficient amount, such that the peel strength of the cured silicone rubber from the molded substrates after curing is enhanced.

A number of test plaques were prepared for testing the PC compositions. The test plaques were molded using the following conditions (63):

The compositions were blended in a twin screw extruder at 260°C and injected into a mold of the size 5 *cm*× 5 *cm* × 0.3 *cm* for unfilled resins and 5 *cm* × 5 *cm* × 2.0 *cm* for fiber-reinforced resins. The blend was injected at 300°C, with an injection speed of 200 $mm\,s^{-1}$ and an injection pressure of 2500 $kp\,cm^{-2}$. The residence time in the mold was about 40 *s*. Then, the plaques were removed and cooled.

Afterwards a liquid silicone rubber composition containing an alkenyl group-containing organopolysiloxane, an organohydrogenpolysiloxane having at least two hydrogen atoms each directly attached to a silicon atom in a molecule, an addition reaction catalyst, selected from platinum and platinum compounds, was applied to one surface of the plaques in a strip of 2 *cm* wide, 4 *cm* long and having a thickness of 0.4 *mm* for plaques without glass fibers and 2.0 *mm* for plaques with glass fiber.

The plaques with the applied silicone rubber composition were cured, cooled and a peel test was done according to ASTM D3330 (67).

The addition of a modifier to the PC improves the peel force by about 30% or greater. The articles exhibit an excellent bonding strength between the substrate PCs that contain the modified poly(dimethyl siloxane)s. The peel strength of the cured silicone rubber from an unfilled substrate after curing at 115°C for 5 *min* may be about 1.80 $kg\,mm^{-1}$ or greater; the substrate has a thickness of 0.3 *mm* (63).

1.3.6 *Urethane-Containing Silane*

Rapid crosslinking of silicone chains on the nanosilica surface during arc discharge has been proposed as an efficient strategy to improve the tracking and erosion resistance of silicone rubber (68).

Urethane-containing silane was synthesized from (3-isocyanatopropyl)triethoxysilane, isopropyl alcohol and allyl alcohol via nucleophilic addition and transetherification (68). The synthesis is shown in Figure 1.19.

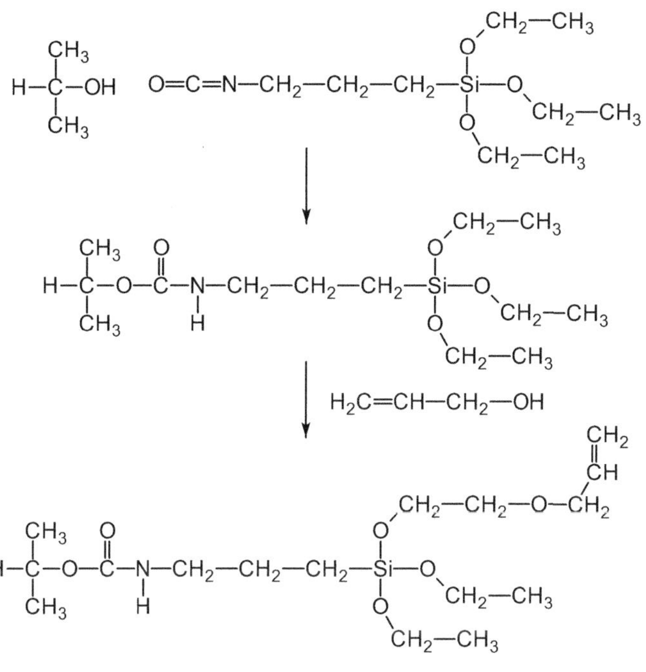

Figure 1.19 Synthesis of urethane-containing silane (68).

The suppression effect and mechanism of urethane-containing silane on the tracking and erosion of addition-cure liquid silicone rubber/silica nanocomposite were investigated.

It was found that urethane-containing silane could significantly enhance the tracking and erosion resistance of addition-cure liquid silicone rubber/silica. When only 2.5 *phr* urethane-containing silane was added, all the addition-cure liquid silicone rubber/silica/urethane-containing silane specimens passed the inclined plane test at 4.5 *kV* and the erosion rate decreased from 3.44% to 0.25%.

This might occur, because urethane-containing silane tightly constrained silicone chains on the silica surface by strong hydrogen-bonding interactions. During the arc discharge, urethane-containing silane at the interphase enhanced the catalytic activity of platinum compound to promote the radical crosslinking of silicone chains absorbed on the silica surface.

Hence, a compact and intact ceramic layer was quickly formed on the addition-cure liquid silicone rubber surface and protected the underlying materials from further arc ablation. These results provide a new method and theoretical basis for the design, preparation and application of the novel high-performance polymer insulating material (68).

1.3.6.1 *Thermal Stability*

A reactive urea-containing silane, (γ-diethylureidopropyl)allyloxyethoxysilane, was synthesized by a transetherification reaction (69). The synthesis is shown in Figure 1.20. The chemical structure was characterized by FTIR and ^{1}H NMR.

The silane was incorporated into addition-cure liquid silicone rubber using a hydrosilylation reaction. The thermal stability of the addition-cure liquid silicone rubber samples was investigated by TG and TG-FTIR.

When the urea-containing silane was incorporated, the temperature of 10% weight loss and 20% weight loss under air atmosphere were increased by 31°C and 60°C in comparison to those of the addition-cure liquid silicone rubber without the urea-containing silane. In addition, the residual weight at 800°C increased from 33.5% to 58.7%.

It was concluded that the enhancement in the thermal stability of the addition-cure liquid silicone rubber samples could be attributed to the decomposition of the urea groups to isocyanic acid, which reacted with hydroxyl groups to inhibit the unzipping depolymerization (69).

1.3.7 *Glass Fiber Fabric*

High silica glass fiber fabric/liquid silicone rubber composite sheets filled with hollow glass microspheres were prepared (70). The effects of the hollow glass microspheres content on the mechanical, thermal

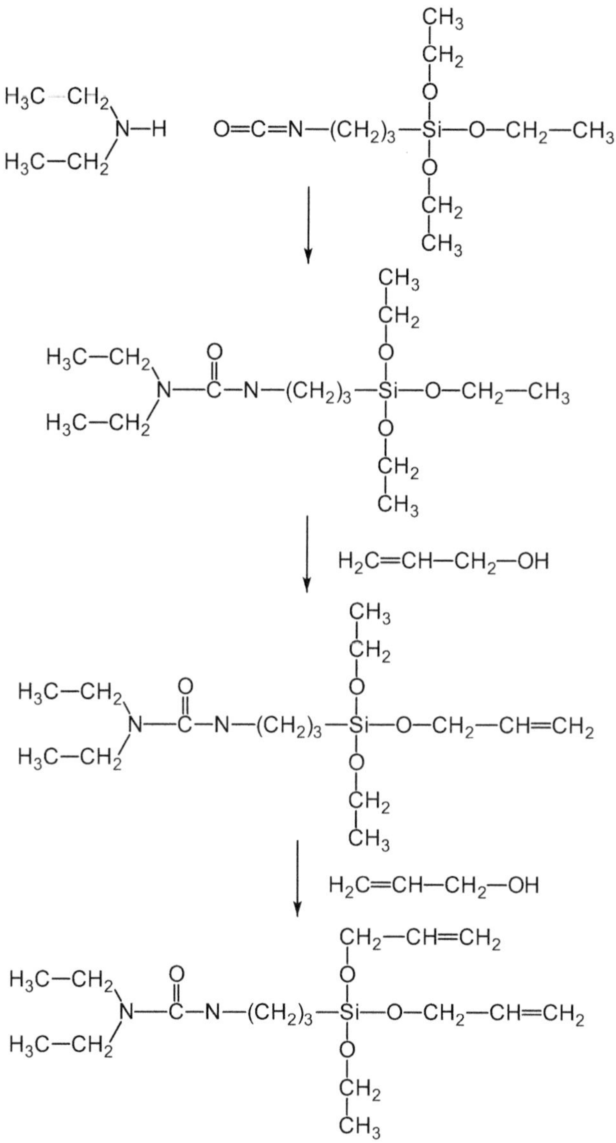

Figure 1.20 Synthesis of (γ-diethylureidopropyl)allyloxyethoxysilane (69).

insulation, and the dynamic mechanical properties of the composite sheets were investigated.

The results showed that the compatibility and interfacial properties between the hollow glass microspheres and the matrix were improved when the hollow glass microspheres were treated with a silane coupling agent, KH550. This is a 3-aminopropyltriethoxysilane coupling agent, c.f. Figure 1.21.

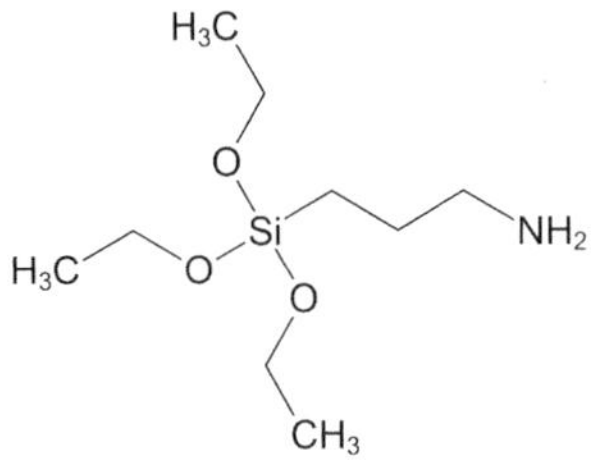

Figure 1.21 3-Aminopropyltriethoxysilane.

A composite in which the hollow glass microspheres' content was 7% by weight, showed the most advantageous mechanical, insulating and damping properties (70).

1.3.8 Foams

1.3.8.1 Cellular Foams

An overview of manufacturing processes used in the development of cellular silicone for a wide variety of applications has been given (71).

The combination of the intrinsic properties of silicone and a foam is considered an attractive solution for many applications. Foaming is very common in hydrosilylation/condensation reactions. This well-known technology leads to homogeneous, elastic, low density and biocompatible foams. However, if the size of the cells remains large, the reactions are sensitive to humidity and the dangerousness of the hydrogen could be an industrial concern. Alternatives for the manufacture of silicone cellular materials are gas foaming, phase separation, emulsion and sacrificial models, and syntactic charges.

In addition, the theories of sorption, diffusion, nucleation and cell growth were detailed to explain the formation of gaseous foam (71).

CO_2 is commonly used to physically foam silicone because of its good solubility. However, the diffusive behavior of CO_2 is high in silicone, as explained by the free volume theory. Silicone-CO_2 foaming is essentially triggered by rapid depressurization, leading to a cell density of around 1×10^9 cells per cm^3 in the best case. Templated foams can be subdivided into emulsion polymerization foams, sacrificial foams, and syntactic foams. These methods are simple because they do not need specific foaming equipment. Also, the pore sizes are tunable as a function of the template sizes (71).

1.3.8.2 Microcellular Crosslinked Silicone Rubber Foams

A series of microcellular silicone rubber/silica/POSS foams were prepared using supercritical carbon dioxide (72).

The effect of the POSS particles on the rheological behavior, vulcanizing, and cellular morphology of the composites was investigated. The results showed that the POSS grafted carboxylic acid group can improve the matrix strength of silicone rubber. POSS grafted carboxylic acid group act as inhibiting agent in the vulcanizing process.

POSS particles play an important role in the microcellular structure formation. When the POSS content was 2.0%, the cell size and cell density can reach 3.77 μm and 7.99×10^9 cells per cm^3, respectively (72).

Influence of Reinforcing Agent on Cellular Morphology. A series of microcellular silicone rubber foams were prepared using supercritical carbon dioxide (73).

A methylvinyl silicone rubber and fumed silica were used. The hydrophobic silica was modified by the hexamethyldisilazane, c.f. Figure 1.22. Dicumyl peroxide was used for vulcanizing. In the compositions, 10 parts of methylvinyl silicone rubber, 6 parts of hydroxy silicone oil, and 1 part of dicumyl peroxide were used, whereas the amount of silica varied from 40 to 70 parts.

The effect of silica as reinforcing agent on the rheological behavior, cellular morphology and nucleation of silicone rubber composites was investigated.

The results indicated that the silica not only acts as reinforcing agent but also plays an important role in the cellular nucleation.

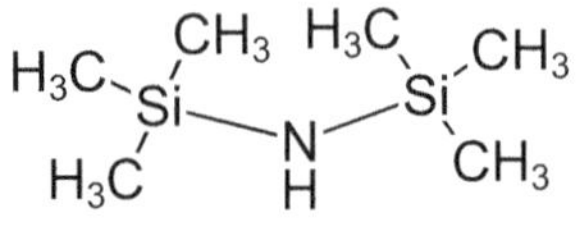

Figure 1.22 Hexamethyldisilazane.

When the content of silica increases from 40 *phr* to 70 *phr*, the range of the calculated surface tension of the silicon rubber composites and the radius of the critical nucleus are 158.95 $nN\,m^{-1}$ to 1,092.74 $nN\,m^{-1}$ and 14.45 *nm* to 99.34 *nm*, respectively.

The aggregated silica shows a good heterogeneous nucleation as the diameter of aggregated silica particles becomes approximate twice. The smallest cell diameter and the highest cell density of the silicone rubber foam can reach 708 *nm* and 1.02×10^{11} cells/cm^3, which indicates that the silicone rubber nanofoams can be obtained by means of the supercritical foaming technology (73).

1.3.9 *Addition Type Liquid Phenyl Silicone Rubber*

A phenyl silicone resin reinforced addition type liquid phenyl silicone rubber was prepared by the vulcanization of vinyl end-capped poly(methylphenyl siloxane), phenyl MT (74) resins, and hydrogen end-capped poly(diphenyl siloxane) under Pt catalysis at 150°C for 4 *h* (75). The effects of the proportion and the vinyl content of phenyl MT resins on the mechanical properties of cured products were investigated. The vulcanizing mechanism of the phenyl silicone rubber with tetraethoxysilane is shown in Figure 1.23.

The thermal stability was explored by TGA. The changes in mechanical properties were also studied before and after irradiation. A phenyl silicone rubber with good performance was obtained when the phenyl MT resin content was 50% to 60% and the vinyl content was at 5.5% to 6.6 wt%. The onset temperature of thermal degradation and the center temperature of thermal degradation were 443.7°C and 502°C, respectively.

When the radiation dose increased from 0 to 300 *KGy*, the tensile strength decreased from 4.1 *MPa* to 2.3 *MPa* and the tearing strength

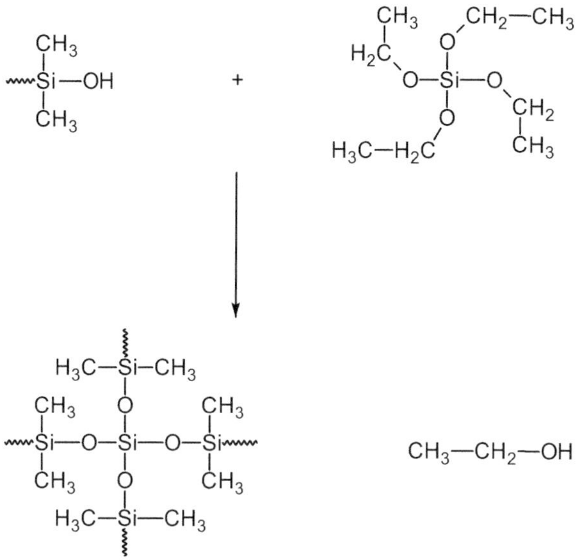

Figure 1.23 Vulcanizing mechanism of phenyl silicone rubber (75).

decreased from 8.9 *MPa* to 5.1 *MPa*. When the radiation dose continued to increase from 300 to 900 *KGy*, the tensile strength increased from 2.3 *MPa* to 6.4 *MPa* and the tearing strength increased from 5.1 *MPa* to 6.5 *MPa*. During the process of radiation, the elongation at break had been kept down from 96% to 52% (75).

A thermal degradation mechanism of room temperature vulcanized phenyl silicone rubber that was vulcanized with different crosslinking agents has been discussed (76).

Room temperature vulcanized phenyl silicone rubber samples were prepared by curing hydroxyl-terminated poly(methyldiphenyl siloxane) via three crosslinking agents, namely, tetraethoxysilane, tetrapropoxysilane, and polysilazane, c.f. Figure 1.24.

The ablation properties of room temperature vulcanized phenyl silicone rubber were studied by the muffle roaster test and FTIR. Then, the thermal stability of the three samples was studied by TG analysis.

In order to explore the thermal degradation mechanism, the room temperature vulcanized phenyl silicone rubber vulcanized by different crosslinking agents was characterized by thermogravimetric

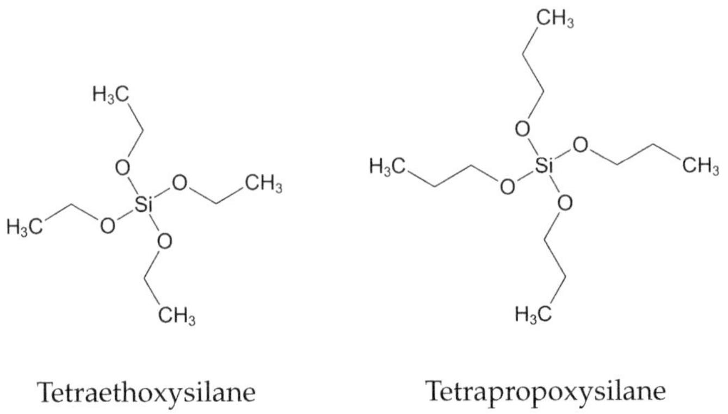

Figure 1.24 Crosslinking agents.

analysis-mass spectrometry (TG-MS) and pyrolysis-gas chromatography mass spectrometry (Py-GC-MS).

The thermal stability of room temperature vulcanized phenyl silicone rubber is related to the amount of the residual Si–OH groups. The residual Si–OH groups initiated the polysiloxane chain degradation via an unzipping mechanism (76).

1.3.10 Organic Foaming Agent

The silicone rubber foam can basically be obtained by compounding an organic foaming agent such as 2,2′-azobisisobutyronitrile, azodicarbonamide, dinitrosopentamethylenetetramine, c.f. Figure 1.25, and a curing agent into a thermosetting millable silicone rubber composition, and then foaming and curing the resultant silicone rubber composition by heating. In order to obtain a silicone rubber foam having excellent characteristics, it is necessary to balance the foaming rate (decomposition rate) of the organic foaming agent and the curing rate of the silicone rubber composition.

If the decomposition rate of the organic foaming agent is lower than the curing rate of the silicone rubber composition, cells become fine but the foaming ratio becomes low, whereas if the curing rate of the silicone rubber composition is lower than the decomposition rate of the organic foaming agent, gas generated by decomposition of the organic foaming agent cannot be sufficiently captured, resulting in a low foaming ratio and an uneven cell structure.

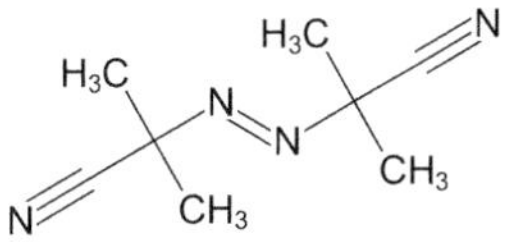

2,2′-Azobisisobutyronitrile

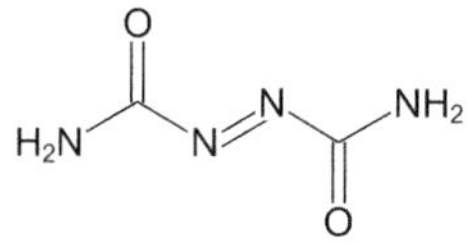

Azodicarbonamide

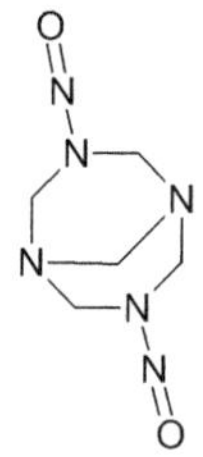

Dinitrosopentamethylenetetramine

Figure 1.25 Organic foaming agents.

In addition, since the organic foaming agent has a high decomposition rate and rapidly generates gas, it is necessary to bring the silicone rubber composition into a predetermined crosslinked state at a time when the organic foaming agent decomposes in order to capture the generated gas.

In crosslinking by peroxide, the crosslinking rate can be adjusted to a certain degree by selection of an organic peroxide and a compounding amount, but it is difficult to strictly adjust the crosslinking density at a time of decomposition of the organic foaming agent, thus resulting in the disadvantage of requiring the compounding of an excessive amount of organic foaming agent or strictly managing process conditions.

A composition for silicone rubber foam may contain:

1. 100 parts by mass of alkenyl group-containing poly(organo siloxane) with a degree of polymerization of 4,000 to 10,000,
2. An amount of poly(organo hydrogen siloxane) having an average of two or more hydrogen atoms bonded to a silicon atom in a molecule so that a molar ratio of the sili-

con-atom-bonded hydrogen atoms to the alkenyl groups in the component 1 is 0.001 to 5,
3. 0.1 to 10 parts by mass of an organic foaming agent with a decomposition temperature of 50–250°C,
4. 5 to 200 parts by mass of silica powder, and
5. A catalyst amount of a platinum-based metal compound catalyst activated with ultraviolet rays.

A manufacturing method of a silicone rubber foam includes (77):

1. Irradiating the composition for silicone rubber foam with ultraviolet rays at a temperature lower than the decomposition temperature of the organic foaming agent to obtain a crosslinked body, and
2. Heat-treating the crosslinked body at a temperature equal to or higher than the decomposition temperature of the organic foaming agent to obtain a silicone rubber foam.

1.3.11 *Foams without Chemical Blowing Agents*

Various methods can be used to prepare foamed polymers (78). Some methods involve the use of a physical or chemical blowing agent. Physical blowing agents are generally volatile liquids that can be used to create voids in a matrix, thereby producing a cellular or foamed material.

Common physical blowing agents include chlorofluorocarbons, hydrochlorofluorocarbons, hydrocarbons and liquid carbon dioxide. Chemical blowing agents expand the foam using one or more chemical reactions that produce a gas. An exemplary chemical blowing agent is powdered titanium hydride, which can be used to make metallic foams by decomposing into titanium and hydrogen gas at elevated temperatures. Other foaming methods include mechanical frothing, extraction of soluble fillers, and use of hollow sphere and/or porous particles (78).

Silicone materials in general have low thermal conductivity, low chemical reactivity, low toxicity, good oxidation stability and good thermal stability. These properties make them particularly useful in thermoacoustic, electrical insulation, and sealing applications that demand high temperature resistance, low smoke generation, and/or high biocompatibility.

One of the potential disadvantages of using blowing agents, frothing, fillers, and hollow or porous particles to create voids in a foam is the potential introduction of impurities into the foam material. This can be problematic in applications where such impurities might cause an adverse chemical or biological reaction. Another limitation faced by some of these methods is the lack of independence between the steps of foaming of the material and the polymerization of the foam material. If these processes occur simultaneously, it can be technically challenging to obtain a desirable cell size and/or obtain a suitable distribution of cell sizes.

It was discovered that a silicone precursor derived from tetraalkoxysilane (Q unit) and trimethylethoxysilane (M unit), known as *MQ resin*, can effectively absorb a significant amount of water to obtain a stable MQ-water cluster. Because the water can be subsequently removed by vaporization, this MQ-water cluster can act as a blowing agent to provide foamed silicone structures. The steps of preparation run as follows (78):

1. Disperse water into the MQ resin to form a stabile cluster.
2. Disperse the MQ-water cluster into the silicone fluid to get an inverted emulsion.
3. Polymerize the silicone emulsion.
4. Foam the emulsion by removing the water.
5. Get the final product.

A micrograph of the produced silicone foam is shown in Figure 1.26.

This micrograph shows the open cells. The cells are separated by polymeric struts. Depending on the relative amount of water dispersed in the MQ-water cluster and the relative amount of MQ-water cluster dispersed in the silicone emulsion, it is possible to tailor the thickness of the polymeric struts. The geometry of the struts can influence mechanical characteristics of the overall foam, including, for example, strength, friability, and pressure required to achieve a compression deflection of 25% (78).

1.3.12 *Epoxy-Silicone Copolymer*

The introduction of an epoxy resin into a liquid silicone rubber not only improves adhesion strength but also the mechanical properties

Figure 1.26 Micrograph of the silicone foam (78).

of the silicone rubber (79).

Two different prepolymers were obtained through an efficient hydrosilylation reaction by the changing reaction time and the catalyst dosage in the presence of poly(oxyethylene) epoxy resin and a hydrogen silicone oil with phenyl groups. The reaction is shown in Figure 1.27.

The chemical structure of the prepolymers was confirmed by FTIR ^{1}H-NMR. Then the prepolymers were incorporated into phenyl-containing silicone rubber at different mass fractions. SEM observations demonstrated that a *sea-island* phase separation occurred in the cured composites.

TGA revealed that the thermal stability of silicone rubber composites improved markedly after modification, the residual yield at 800°C achieved 32.69% in the air atmosphere when 20 *phr* hydrogen silicone oil with phenyl groups were introduced.

The studies of the mechanical properties showed that the tensile strength and the shear strength of composites increased gradually with increasing prepolymers.

These results provide very valuable information for understand-

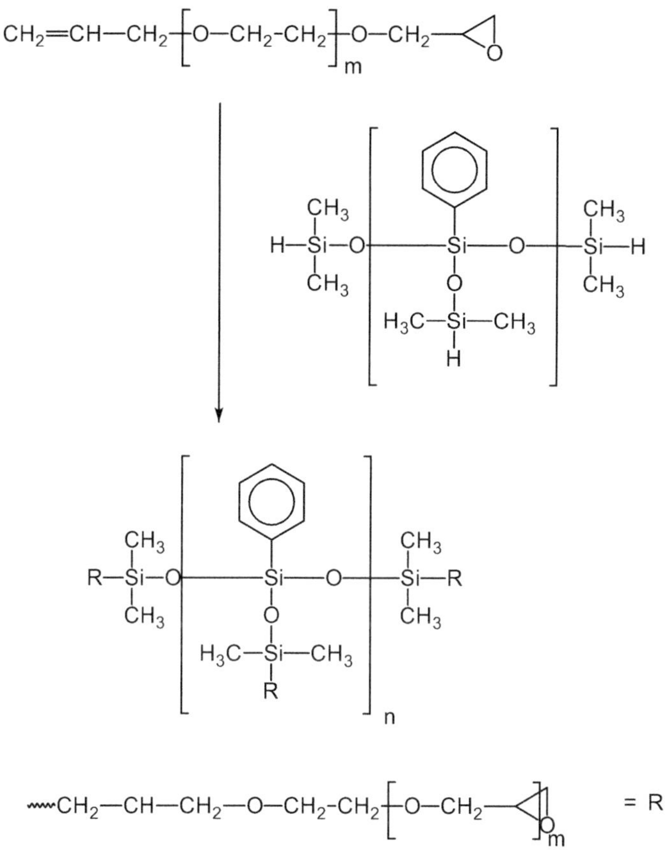

Figure 1.27 Hydrosilylation reaction (79).

ing the hydrosilylation reaction and the effects of the molecular structures on the performance of silicone rubber systems (79).

1.3.12.1 *Silicone Defoamer*

Many petroleum refineries utilize a delayed coking unit to process the residual oils (80). In a delayed coking, overhead vapors from the coking drum pass through a fractionator where various fractions are separated. One of the fractions is a gasoline boiling range stream.

This stream, commonly referred to as coker gasoline or coker naphtha, is generally a relatively low octane stream, not suitable for use as an automotive fuel without upgrading. With the limitations on octane-enhancing additives imposed by environmental concerns, it has become desirable to process coker gasoline naphtha in an octane-enhancing unit such as a catalytic reformer. Sulfur compounds in coker gasoline are detrimental to catalytic reforming catalysts, and accordingly it has become commonplace to treat coker gasoline in a catalytic desulfurizer prior to processing it in a catalytic reformer. These steps of delayed coking, desulfurization and reforming are well-developed commercial processes which are widely used in the refining industry. Hydrodesulfurization (HDS) catalysts can be undesirably poisoned by silicon. Silicon can also be the poison reforming catalyst, which is even more expensive.

In the delayed coking process, quantities of foam are often produced which are undesirable, as the foam diminishes the efficiency of the delayed coking process. It is customary to add silicone defoamers, such as poly(dimethyl siloxane)s, to delayed cokers to control the undesirable foaming (81). Thus, one of the recurring concerns in the coker is contamination of the products with silicon species that can eventually poison a hydrotreater catalyst and cause the expensive catalyst to be replaced prematurely.

Since the silicone-based defoamer used in the coke drum is believed to be a major source of the silicon contamination, considerable effort has been expended to minimize or eliminate the use of silicone oil from the unit (80). The silicone oil acts both as a defoamer to knock down foam in the drum when the silicone oil is first injected and also acts as an antifoam agent to prevent the subsequent build up of foam. Due to silicone oil's unique surface properties and thermal stability, there has not yet been a commercially viable non-silicone defoamer replacement.

It has been found that crosslinked PDMS resins are useful defoamers and antifoamers for hydrocarbon-containing liquids, such as delayed coker feedstocks and feedstocks to preflash and atmospheric towers (80).

These PDMS resins are crosslinked with either an alkyl poly(silicate) or siloxane. The crosslinked PDMS resins may be used alone or together with linear PDMS, and are typically blended with a carrier such as kerosene for easier handling. Importantly, the use of crosslinked branched PDMS resins permit less total amount of poly(siloxane)s to be used, which reduces the silicon carryover in coker products and reduces poisoning of downstream catalysts (80).

References

1. K.E. Polmanteer, *Rubber Chemistry and Technology*, Vol. 61, p. 470, 1988.
2. J. Ackermann and V. Damrath, *Chemie in unserer Zeit*, Vol. 23, p. 86, 1989.
3. A.F. Jacobine and S.T. Nakos, Photopolymerizable silicone monomers, oligomers, and resins in S.P. Pappas, ed., *Radiation Curing: Science and Technology*, pp. 181–240. Springer US, Boston, MA, 1992.
4. R.B. Allen, P. Kochs, and G. Chandra, Industrial organosilicone materials, their environmental entry and their predicted fate in G. Chandra, ed., *Organosilicon Materials*, Vol. 3H of *The Handbook of Environmental Chemistry*, chapter 1, pp. 1–26. Springer, Berlin New York, 1997.
5. D.E. Katsoulis, R.G. Schmidt, and G.A. Zank, Siloxanes and silicones (advances in silicone technologies 2000-15) in V.Y. Lee, ed., *Organosilicon Compounds*, chapter 7, pp. 301–322. Academic Press, 2017.
6. V. Lee, ed., *Organosilicon Compounds*, Academic Press, London, 2017.
7. M.T. Nazir, B.T. Phung, S. Yu, and S. Li, *IEEE Transactions on Dielectrics and Electrical Insulation*, Vol. 25, p. 2076, Dec 2018.
8. IEC, Electrical insulating materials used under severe ambient conditions – test methods for evaluating resistance to tracking and erosion, IEC 60587, International Electrotechnical Commission, Geneva, Switzerland, 2007.
9. N. Yan, Z. Jia, and X. Zou, *IEEE Transactions on Dielectrics and Electrical Insulation*, Vol. 25, p. 162, February 2018.
10. IEC, Dry, solid insulating materials – resistance test to high-voltage, low-current arc discharges, IEC 61621, International Electrotechnical Commission, Geneva, Switzerland, 1997.

11. S. Ansorge, F. Schmuck, and K.O. Papailiou, *IEEE Transactions on Dielectrics and Electrical Insulation*, Vol. 22, p. 979, April 2015.
12. C. Geng, Q. Zhang, W. Lei, F. Yu, and A. Lu, *Journal of Applied Polymer Science*, Vol. 134, p. 45544, 2017.
13. B.M. Maher, J. Rezaali, K. Jalili, and F. Abbasi, *Rubber Chemistry and Technology*, Vol. 90, p. 108, 2017.
14. H. Ghaleh, K. Jalili, B.M. Maher, R. Rahbarghazi, M. Mehrjoo, S. Bonakdar, and F. Abbasi, *European Polymer Journal*, Vol. 106, p. 305 , 2018.
15. D. Blanc, C. Maliverney, and R. Ferhat, Article having antifouling properties and intended to be used in aquatic applications, particularly marine applications, US Patent Application 20 110 305 911, assigned to Bluestar Silicones France, Lyon Cedex (FR), December 15, 2011.
16. A. Milne, Anti-fouling marine compositions, US Patent 4 025 693, assigned to The International Paint Co., Ltd. (London, ENG), May 24, 1977.
17. A. Acton, ed., *Siloxanes – Advances in Research and Applcation*, p. 95. Scholary Editions, Atlanta, Georgia, 2013.
18. B.M. Maher, J. Rezaali, H. Ghaleh, S. Fallahnejad, F. Abbasi, and A.A. Babaluo, *Colloid and Polymer Science*, Vol. 295, p. 1595, September 2017.
19. C. Baumgart, D. Weiß, and V. Altstädt, *Polymer Engineering & Science*, Vol. 56, p. 849, 2016.
20. H. Sirin, M. Kodal, B. Karaagac, and G. Ozkoc, *Composites Part B: Engineering*, Vol. 98, p. 370, 2016.
21. V. Seitz, K. Arzt, S. Mahnel, C. Rapp, S. Schwaminger, M. Hoffstetter, and E. Wintermantel, *International Journal of Adhesion and Adhesives*, Vol. 66, p. 65, 2016.
22. M. Shimakawa, Silicone rubber composition, method of producing the same, and silicone rubber extrudate, US Patent Application 20 180 291 160, assigned to Momentive Performance Materials Japan LLC, Tokyo (JP), October 11, 2018.
23. K. Han, G. Jo, and B. Kim, Silicone composition and a pressure sensitive adhesive film having a pressure sensitive adhesive layer made from the composition, US Patent Application 20 170 233 612, assigned to Dow Silicones Corp., August 17, 2017.
24. S. Aoki and A. Yamamoto, Silicone composition and a pressure sensitive adhesive film having a pressure sensitive adhesive layer made from the composition, US Patent 8 057 909, assigned to Shin-Etsu Chemical Co., Ltd. (Tokyo, JP), November 15, 2011.
25. T. Morita, Y. Nimura, R. Yamaoka, S. Fukagawa, and T. Seki, Silicone rubber composition and silicone rubber crosslinked body, US

Patent Application 20 180 208 767, assigned to Sumitomo Riko Company Limited, Komaki-shi (JP) and SumiRiko Fine Elastomer, Ltd., Ageo-shi (JP), July 26, 2018.
26. W. Chen, X. Zeng, X. Lai, H. Li, W. Fang, and T. Liu, *Thermochimica Acta*, Vol. 632, p. 1, 2016.
27. S. Azechi, Y. Taira, and T. Yoshida, Addition curing type liquid silicone rubber compositions, US Patent 6 331 588, assigned to Shin-Etsu Chemical Co., Ltd. (Tokyo, JP), December 18, 2001.
28. M. Scholz and J. Meyer, Silicone rubber, US Patent 9 708 458, assigned to Evonik Degussa GmbH (Essen, DE), July 18, 2017.
29. ASTM Subcommittee, Standard test method for silica-*n*-dibutyl phthalate absorption number, ASTM D6854, ASTM International, West Conshohocken PA, 2003.
30. ASTM Subcommittee D24.11, Standard test method for carbon black--oil absorption number of compressed sample (COAN), ASTM D3493, ASTM International, West Conshohocken PA, 2018.
31. H.Y. Jin, Y.F. Li, S.C. Nie, P. Zhang, N.K. Gao, and W. Li, *Materiali in Tehnologije*, Vol. 51, p. 783, 2017.
32. Y. Li, H. Jin, S. Nie, C. Tong, and N. Gao, *AIP Advances*, Vol. 8, p. 015313, 2018.
33. Y. Li, H. Jin, S. Nie, P. Zhang, and N. Gao, *Applied Physics Letters*, Vol. 110, p. 201602, 2017.
34. J. Song, C. Chen, and Y. Zhang, *Composites Part A: Applied Science and Manufacturing*, Vol. 105, p. 1, 2018.
35. J. Wang, D. Zhao, X. Zou, L. Mao, and L. Shi, *Journal of Materials Science: Materials in Electronics*, Vol. 28, p. 12984, 2017.
36. Y. Xue, X. Li, H. Wang, F. Zhao, D. Zhang, and Y. Chen, *Materials & Design*, Vol. 165, p. 107580, 2019.
37. A. Lendlein, ed., *Shape-Memory Polymers*, number 226 in Advances in Polymer Science, Springer, Heidelberg, New York, 2010.
38. J. Leng and S. Du, eds., *Shape-Memory Polymers and Multifunctional Composites*, CRC Press, Boca Raton, FL, 2010.
39. M. Pantoja, Z. Lin, M. Cakmak, and K.A. Cavicchi, *Journal of Polymer Science Part B: Polymer Physics*, Vol. 56, p. 673, 2018.
40. Y. Guo, *Investigation of Silicone Rubber Blends and their Shape Memory Properties*. PhD thesis, University of Akron, Akron, Ohio, 2018.
41. J. Hu, ed., *Advances in Shape Memory Polymers*, number 146 in Woodhead Publishing Series in Textiles, Woodhead Publishing, Cambridge, UK, 2013.
42. Q. Meng, J. Hu, Y. Zhu, J. Lu, and Y. Liu, *Journal of Applied Polymer Science*, Vol. 106, p. 2515, 2007.
43. L. Xu, Thermoplastic elastomer compounds exhibiting shape memory via thermo-mechanical action, US Patent 9 290 652, assigned to PolyOne Corporation (Avon Lake, OH), March 22, 2016.

44. K. Tsuji, K. Yamada, H. Kizaki, and N. Matsumoto, Thermally conductive silicone grease composition, US Patent 9 321 950, assigned to Shin-Etsu Chemical Co., Ltd. (Tokyo, JP), April 26, 2016.
45. X. Luo and D.D.L. Chung, *Journal of Intelligent Material Systems and Structures*, Vol. 8, p. 389, 1997.
46. N.A. Kotov, *Nature*, Vol. 442, p. 254, July 2006.
47. A. Yu, P. Ramesh, M.E. Itkis, E. Bekyarova, and R.C. Haddon, *The Journal of Physical Chemistry C*, Vol. 111, p. 7565, 2007.
48. V. Causin, C. Marega, A. Marigo, G. Ferrara, and A. Ferraro, *European Polymer Journal*, Vol. 42, p. 3153, 2006.
49. S. Ganguli, A.K. Roy, and D.P. Anderson, *Carbon*, Vol. 46, p. 806, 2008.
50. S. Bittolo Bon and L. Valentini, *arXiv preprint*, Vol. 8, 2015.
51. Y. Zhang, X. Zeng, X. Lai, and H. Li, *RSC Advances*, Vol. 8, p. 111, 2018.
52. UL LLC, Standard for tests for flammability of plastic materials for parts in devices and appliances, Online, electronic: https://standardscatalog.ul.com/standards/en/standard_94_6 UL 94, Underwriters Laboratories of the United States, Northbrook, Illinois, 2015.
53. P. Rybiński, B. Syrek, D. Bradło, and W. Żukowski, *Materials*, Vol. 11, p. 1298, 2018.
54. L.A. Bloomfield, *arXiv preprint arXiv:1801.09253*, Vol. 1, p. 1, 2018.
55. W. Noll, ed., *Chemistry and Technology of Silicones*, Academic Press, Oxford, 1968.
56. R.W. Boot, Process for preparing organo-polysiloxane compositions, US Patent 3 213 048, assigned to General Electric, October 19, 1965.
57. L.A. Bloomfield, Viscoelastic silicon rubber compositions, US Patent 9 550 864, assigned to University of Virginia Patent Foundation (Charlottesville, VA), January 24, 2017.
58. S. Rubinsztajn, *Journal of Inorganic and Organometallic Polymers and Materials*, Vol. 24, p. 1092, November 2014.
59. X. Zhao, C. Zang, Y. Sun, K. Liu, Y. Wen, and Q. Jiao, *Journal of Materials Science*, Vol. 53, p. 1167, January 2018.
60. J. Sun, J. Kong, and C. He, *Journal of Applied Polymer Science*, Vol. 136, p. 46996, 2019.
61. X. Li, Q. Wu, M. Zheng, Q. Li, S. Wang, and C. Zhang, *Cellulose*, Vol. 25, p. 473, January 2018.
62. T. Wu, X. Lai, F. Liu, H. Li, and X. Zeng, *Applied Surface Science*, Vol. 459, p. 483, 2018.
63. K.-l.L. Cheng, C.W. Hung, C.T. van Nuffel, and C.-W.W. Chang, Molded structures of polycarbonate based substrates over molded with silicone rubbers, US Patent Application 20 190 031 916, assigned to Trinseo Europe GmbH, January 31, 2019.

64. H. Fujiki, S. Shudo, A. Matsuda, N. Ogawa, and T. Takata, Polycarbonate resin/silicone rubber integrally molded article and method for making, US Patent 5 366 805, assigned to Shin-Etsu Chemical Company, Ltd. (Tokyo, JP), November 22, 1994.
65. H. Fujiki and S. Shudo, Thermoplastic resin compositions for use in integral molding with silicone rubber and integrally molded parts, US Patent 5 792 812, assigned to Shin-Etsu Chemical Co., Ltd. (Tokyo, JP), August 11, 1998.
66. H. Fujiki and S. Shudo, Thermoplastic resin compositions for use in integral molding with silicone rubber and integrally molded parts, EP Patent 0 722 990, assigned to Shin-Etsu Chemical Co. Ltd., February 12, 1997.
67. ASTM Subcommittee D10.14, Standard test method for peel adhesion of pressure-sensitive tape, ASTM D3330, ASTM International, West Conshohocken PA, 2018.
68. Y. Zhang, X. Zeng, X. Lai, H. Li, and X. Huang, *Polymer Testing*, Vol. 69, p. 16, 2018.
69. W. Fang, X. Zeng, X. Lai, H. Li, W. Chen, and Y. Zhang, *Thermochimica Acta*, Vol. 605, p. 28, 2015.
70. S. Chen, Y. Qin, J. Song, and B. Wang, *Polimery*, Vol. 63, 2018.
71. T. Métivier and P. Cassagnau, *Journal of Cellular Plastics*, Vol. 55, p. 151, 2019.
72. T. Liu, Y. Lei, F. Zhang, S. Guo, and S. Luo, *Polymer-Plastics Technology and Engineering*, Vol. 57, p. 1623, 2018.
73. B. Xiang, Z. Deng, F. Zhang, N. Wen, Y. Lei, T. Liu, and S. Luo, *Polymer Engineering & Science*, Vol. 59, p. 5, 2019.
74. The Dow Chemical Company, Fascinating silicones, Online, electronic: https://consumer.dow.com/en-us/fascinating-silicones.html, Dow Chemical Company, Midland (Michigan), USA, 2018.
75. D. Peng, Q.H. Mu, S. Zhang, J.H. Li, and F. Wang, Synthesis and properties of phenyl silicone resin reinforced addition type liquid phenyl silicone rubber, in *International Symposium on Advanced Material Research II*, Vol. 926 of *Materials Science Forum*, pp. 39–44. Trans Tech Publications, August 2018.
76. C. He, B. Li, Y. Ren, W. Lu, Y. Zeng, W. He, and A. Feng, *Materials*, Vol. 12, p. 88, December 2018.
77. M. Shimakawa, Composition for silicone rubber foam, manufacturing method of silicone rubber foam, and silicone rubber foam, US Patent 9 340 655, assigned to Momentive Performance Materials Japan LLC (Tokyo, JP), May 17, 2016.
78. J.J. Liu, P. Wu, F. Bai, and D.A. Ylitalo, Silicone compositions and related methods, US Patent 9 540 494, assigned to 3M Innovative Properties Company (Saint Paul, MN), January 10, 2017.

79. B. Zhang, R. Li, J. Luo, Y. Chen, H. Zou, and M. Liang, *Polymer Bulletin*, Vol. 75, p. 2105, May 2018.
80. L.N. Kremer, Silicone defoamer to better control hydrocarbon foam and reduce silicon content of liquid hydrocarbon products, US Patent 7 427 350, assigned to Baker Hughes Incorporated (Houston, TX), September 23, 2008.
81. J.A. Hyde, Silicone oil antifoam, US Patent 3 700 587, assigned to Nalco Company LLC, October 24, 1972.

2

Methods

2.1 Special Curing Methods

2.1.1 Precision Casting

Wax patterns can be used in precision casting (1). A low-cost approach to a rapid fabrication of customized wax patterns with complex geometry is an important research issue.

A wax filament with low melting point has been developed for fabricating wax patterns using the additive manufacturing technology (1). The optimal process parameters for producing wax pattern were investigated using the Taguchi design method. The results showed that the most important control factor affecting the form accuracy of the fabricated wax patterns is the flow of nozzle, followed by print speed. In detail, the process runs as (1):

1. Making a filament,
2. Making a rubber silicone mold,
3. Wax injection molding for a wax filament,
4. Additive manufacturing,
5. Testing with Taguchi design method,
6. Conformal cooling channels,
7. Injection molding with conformal cooling channels.

A cylindrical rod with a dimension of 10 *mm* and 10 *mm* in height was designed with the Pro/ENGINEER CAD software and used as a master model. The edge of the printed cylindrical rod has an obvious stair-stepping effect that can be used for evaluating the form accuracy of the wax patterns.

The form accuracy of the fabricated wax patterns has a lower impact for the nozzle temperature and bed temperature. The optimal process parameters for producing wax patterns are nozzle temperature of 64°C, a print speed of 60 $mm\,s^{-1}$, a bed temperature of 40°C, and flow of nozzle of 100%. An injection molding tooling was fabricated by wax conformal cooling channels fabricated by the optimal process parameters (1).

2.1.2 *Hybrid Additive Manufacturing*

The feasibility of a hybrid additive manufacturing (AM) method combining material extrusion and powder bed binder jetting (PBBJ) techniques for the fabrication of structures made of silicone, i.e., poly(siloxane) was investigated (2).

A full factorial experimental design was conducted to maximize the geometrical accuracy of the parts. The rheological and morphological properties of the silicone powders, the thermal characteristics of the liquid silicone binder, and mechanical characterization of the additively manufactured parts are reported. Using this hybrid AM method, porous cylindrical structures of around 5 *mm* diameter and 3 *mm* height with potential applications in the biomedical industry were additively manufactured.

The final structures were composed of ca. 60% silicone powder, 30% silicone binder, and less than 10% air voids. These three phases were distributed throughout the structure in a non-uniform fashion (2).

2.1.3 *Peroxide Curing*

Silicone rubber compositions that are cured by an organic peroxide curing agent may be used in various applications (3). They are very stable at high and low temperatures, and may exhibit various hardness values. In addition, these compositions have the advantages of excellent chemical stability, water repellency, electrical properties and compression resistance.

The silicone rubber composition may be prepared by mixing a high-viscosity silicone gum containing a vinyl group in the molecule with a reinforcing filler, such as fumed silica, and a processing aid to prepare a base compound, dispersing an organic peroxide curing agent in the base compound, and heating the base compound having the curing agent dispersed therein to yield a desired rubber compound.

The silicone rubber composition may contain a poly(dimethylvinyl siloxane), a silica filler, a poly(methylvinyl siloxane) oil, a hydrogen-modified poly(organo siloxane), a platinum-based catalyst, and a poly(methyl hydrogen siloxane) vulcanizing agent.

The silica filler is included for the purpose of increasing the mechanical strength of the composition. The silica filler may be amorphous or crystalline.

The silica filler may have an average particle size ranging from about 5 nm to about 30 μm and a specific gravity ranging from about 1.5 $g\,cm^{-3}$ to about 2.5 $g\,cm^{-3}$. In this range, the silica filler may be easily dispersed so that the workability and strength of the composition may be improved.

The poly(methylvinyl siloxane) oil is included for the purpose of ensuring the recovery of the composition from compressive deformation. The poly(methylvinyl siloxane) oil may contain a vinyl group at both ends or the side chain. The poly(methylvinyl siloxane) oil has a vinyl content ranging from about 1 $mol\%$ to about 20 $mol\%$. In this vinyl content range, the composition has an excellent ability to recover from compressive deformation.

The hydrogen-modified poly(organo siloxane) is included for the purpose of ensuring the anti-yellowing properties, resilience and mechanical properties of the first cured product even after second curing. The hydrogen-modified poly(organo siloxane) may contain a vinyl group at both ends and hydrogen at the side chain, or may contain a vinyl group at the side chain and hydrogen at both ends.

The platinum-based catalyst is included for the purpose of promoting the curing reaction or the hydrosilylation reaction of the silicone rubber composition. The platinum-based catalyst that is used may be a typical platinum-based catalyst. For example, the platinum-based catalyst may contain one or more compound, selected from fine platinum powder, platinum black, chloroplatinate, alcohol modified chloroplatinate, chloroplatinate-olefin complexes, chloroplatinate-alkenyl siloxane complexes, and chloroplatinate-divinyltetramethyldisiloxane complexes.

The poly(methylhydrogensiloxane) vulcanizing agent is included for the purpose of vulcanization, i.e., crosslinking. The poly(methylhydrogensiloxane) vulcanizing agent may be an organopolysiloxane structure containing a hydrogen atom at both ends or the side chain. For example, the poly(methylhydrogensiloxane) vulcanizing agent may be selected from methylhydrogensiloxanes blocked with a trimethylsiloxy group at both ends, a copolymer of a dimethyl siloxane blocked with a trimethylsiloxy group at both ends and methylhydrogensiloxane, a copolymer of a dimethyl siloxane blocked with a dimethylhydrogensiloxy group at both ends and methylhydrogensiloxane, and tetramethyltetrahydrogen cyclotetrasiloxane.

A plasticizer may be included for the purpose of increasing the compatibility and dispersibility of the hydrophilic silica component and the hydrophobic siloxane component and to improve the storage stability and the moldability of the silicone rubber composition, and also to enhance the physical properties of the silicone rubber. The plasticizer may comprise a poly(dimethyl siloxane) plasticizer, for example, poly(dimethyl siloxane) containing a hydroxyl group at both ends.

A release agent may be included for the purpose of ensuring the release of the silicone rubber composition from a mold. The release agent may be selected from organic fatty acids and their metal salts, e.g., stearic acid, zinc stearate, calcium stearate, and metal salts thereof.

A curing retarder may be included for the purpose of controlling the curing rate of the silicone rubber composition. Curing retarder compounds are collected in Table 2.1 and some compounds are shown in Figure 2.1.

When a silicone rubber is cured by an addition reaction in the presence of a platinum catalyst, yellowing may be eliminated. However, when a silicone rubber product is prepared using fumed silica and a curing method that is performed by an addition reaction in the presence of a platinum catalyst, the resilience of the silicone rubber product is significantly lower than that of a silicone rubber product prepared using precipitated silica and the organic peroxide curing method (3).

A typical preparation procedure runs as follows (3):

Preparation 2–1: First, 100 parts by weight of poly(dimethylvinyl siloxane) was fed into a kneader (5 *l*, Battaggion, Italy) and was uniformly dis-

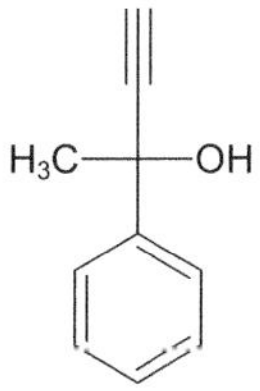

2-Phenyl-3-butyn-2-ol

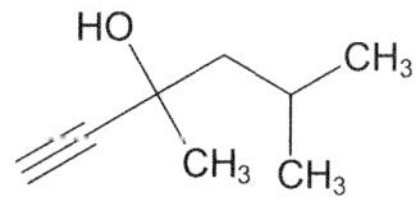

3,5-Dimethyl-1-hexyn-3-ol

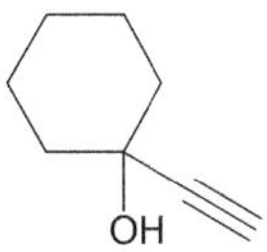

1-Ethynylcyclohexanol

1,5-Hexadiyne

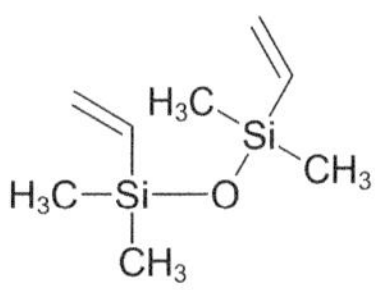

1,3-Divinyltetramethyldisiloxane

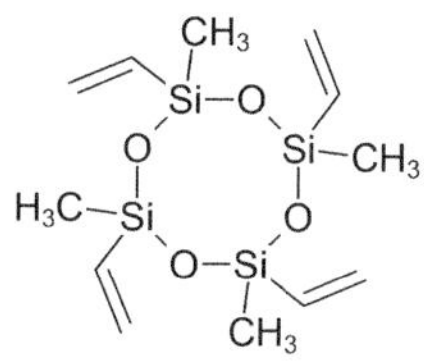

1,3,5,7-Tetravinyltetramethyl-
cyclotetrasiloxane

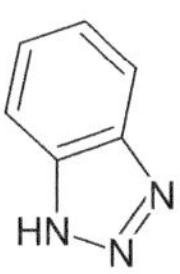

1,2,3-Benzotriazole

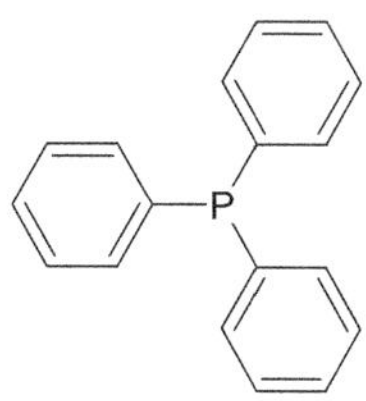

Triphenylphosphine

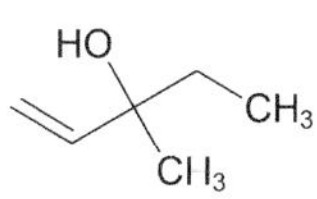

3-Methyl-1-penten-3-ol

Figure 2.1 Curing retarders.

Table 2.1 Curing retarder compounds (3).

Compound	Compound
2-Methyl-3-butyn-2-ol	2-Phenyl-3-butyn-2-ol
3,5-Dimethyl-1-hexyn-3-ol	1-Ethynylcyclohexanol
1,5-Hexadiyne	1,6-Heptadiyne
3,5-Dimethyl-1-hexen-1-yne	3-Ethyl-3-buten-1-yne
3-Phenyl-3-buten-1-yne	Tributylamine
Tetramethylethylenediamine	Benzotriazole
Triphenylphosphine	3-Methyl-1-penten-3-ol
1,3-Divinyltetramethyldisiloxane	
Methyltris(3-methyl-1-butyn-3-oxy)silane	
1,3,5,7-Tetravinyltetramethylcyclotetrasiloxane	
1,3-Divinyl-1,3-diphenyldimethyldisiloxane	

persed. Then, silica filler and poly(methylvinyl siloxane) oil were added to and mixed with the poly(dimethylvinyl siloxane). To increase the dispersing effect, silica filler and poly(methylvinyl siloxane) oil were added in two portions and mixed with the poly(dimethylvinyl siloxane) to prepare a first mixture. The first mixture was heated for 2 *h* while the temperature was elevated from 80°C to 160°C.

Next, the first mixture was depressurized under the conditions of 160°C and 0.8 *bar* for 2 *h* in order to remove water and low-molecular-weight siloxane compound, which are reaction by-products produced in the kneader.

Next, the internal temperature of the kneader chamber was lowered to 60°C, and then hydrogen-modified poly(organo siloxane) was feeded into the kneader. Following this, platinum-based catalyst, plasticizer, release agent and curing retarder were fed into the kneader and dispersed using a two-roll mill for 5 *min*. Then, the poly(methylhydrogensiloxane) vulcanizing agent was fed into the kneader and uniformly dispersed for 5 *min*, thereby preparing a second mixture. The prepared second mixture was subjected to molding and first curing using a hot press at 140°C for 10 *min*, thereby preparing a first cured product (sheet) having a size of 15 *cm* width and 15 *cm* length with 2 *mm* thickness.

2.1.4 *Ultraviolet Curing*

2.1.4.1 *Mesoporous Silica*

Mesoporous silica was used as an inorganic filler material to improve the mechanical strength of silicone rubbers (4).

Mesoporous silica particles have abundant nanoscale pores and a large pore volume. They can be prepared through the spontaneous self-assembly of surfactants (5–8).

The mesoporous silica particles are more effective in reducing the coefficient of thermal expansion and hardening silicone rubber composites in comparison to commercially available nonporous silica particles (4).

An ultraviolet curing type silicone rubber was used for the preparation of mesoporous silica composites using a simple single-step method. From an industrial viewpoint, simplifying the fabrication processes is critical. The thermal stability and mechanical strength were examined in detail in order to show the effectiveness of mesoporous silica particles as filler materials (4).

2.1.4.2 Zeolite

Zeolite, which has regular pores, can adsorb various substances in the pores (9). Accordingly, zeolite is used in a wide variety of applications such as deodorant and hygroscopic agent as well as water softener and separation membrane.

A UV-curable silicone rubber composition has been presented (9). The composition does not undergo curing failure, foaming, and other undesirable conditions even if a water-containing inorganic filler such as zeolite is added.

Preferably, the organohydrogenpolysiloxane used is one which is liquid with the viscosity at 25°C of 1 *mPa s* to about 300 *mPa s*. An example for the preparation of such a composition runs as follows (9):

Preparation 2–2: First, 40 parts of a straight chain organopolysiloxane raw rubber having an average degree of polymerization of about 6,000 containing 99.850% by mole of a dimethyl siloxane unit, 0.125% by mole of methylvinyl siloxane unit as the diorgano siloxane unit constituting the backbone, and 0.025% by mole of dimethylvinylsiloxy group as the molecular chain end group, and 60 parts of high silica zeolite HiSiv 3000 having a water content of 1.8% by weight (weight loss by heat treatment at 200°C for 5 *min*) were homogeneously kneaded in a kneader at room temperature at 25°C to prepare a compound.

Next, 1.4 parts of methylhydrogenpolysiloxane having Si–H group in the side chain (dimethyl siloxane-methylhydrogensiloxane copolymer having opposite ends of the molecular chain endcapped with trimethylsiloxy group having a degree of polymerization of about 80 and a Si–H group content of 0.0055 $mol\, g^{-1}$) as the crosslinking agent and 0.3 part of 0.16% toluene solution of (methylcyclopentadienyl)trimethyl platinum complex as the photoactive platinum catalyst were added to 100 parts of the compound as described above. The mixture was homogeneously kneaded to prepare a silicone rubber composition. An uncured sheet having a thickness of 2 mm was prepared from this silicone rubber composition, and this sheet was exposed to a UV beam of 5,000 $mJ\, cm^{-2}$ to prepare a test sheet.

The silicone rubber composition can be readily cured by UV irradiation. Exemplary UV irradiation sources include a low pressure mercury lamp, high pressure mercury lamp, ultrahigh pressure mercury lamp, xenon lamp, xenon-mercury lamp, and metal halide lamp. The preferred source is a high pressure mercury lamp in view of availability, which may be either an ozone-generating mercury lamp or ozone-free mercury lamp. In view of the association of the high energy wavelength UV radiation, ozone-generating mercury lamp is preferred (9).

2.1.5 *Addition-Curable Compositions*

An addition-curable liquid silicone rubber composition has been developed, which suppresses thickening over time even in cases where a silicone raw rubber is added into a liquid silicone rubber composition (10). The composition shows an excellent storage stability.

It yields a silicone rubber that has a low hardness after being cured. The addition-curable liquid silicone rubber composition contains (10):

1. An alkenyl group-containing organopolysiloxane having an average polymerization degree of 1,500 or less, which is in a liquid state at room temperature,
2. An organopolysiloxane having an average polymerization degree of 2,000 or more, which is in the form of a raw rubber at room temperature,
3. An organohydrogen polysiloxane,
4. A fumed silica having a BET specific surface area of 130 $m^2\, g^{-1}$ or more,

5. An organosilane and/or organopolysiloxane having a silanol group, and
6. An addition reaction catalyst.

This addition-curable liquid silicone rubber composition provides a low-hardness silicone rubber which has a hardness (Durometer A) of 5–40 after being cured (10).

A specific example is (10):

Preparation 2–3: Sixty parts of dimethylpolysiloxane (A1) capped at both ends with dimethylvinylsiloxy groups and having an average degree of polymerization of 750, 40 parts of fumed silica (D1) having a BET specific surface area of 300 $m^2\ g^{-1}$ (Aerosil 300, from Nippon Aerosil Co., Ltd.), 8 parts of hexamethyldisilazane and 2.0 parts of water were mixed together for 30 *min* at room temperature, after which the temperature was raised to 150°C and stirring was continued for 3 *h*, followed by cooling, thereby giving a silicone rubber base.

Fifty parts of the dimethylpolysiloxane (A1), 20 parts of raw dimethylpolysiloxane rubber (B1) capped on both ends with trimethylsiloxy groups and having an average degree of polymerization of 8,000, and 0.5 part of dimethylpolysiloxane (E1) capped on both ends with dimethylhydroxysiloxy groups and having an average degree of polymerization of 10 were placed in 100 parts of this silicone rubber base and stirring was continued for 30 minutes. Next, 0.86 part of methylhydrogenpolysiloxane (C1) capped on both ends with trimethylsiloxy groups and having pendant SiH groups (degree of polymerization, 27; SiH group content, 0.0069 $mol\ g^{-1}$) as a crosslinking agent and 0.05 part of 1-ethynylcyclohexanol as a reaction regulator were added and stirring was continued for 15 *min*, giving a silicone rubber mixture.

Immediately following production, this silicone rubber mixture was stored at 70°C for 7 *d*, after which the viscosity at 25°C and a shear rate of 1 s^{-1} was measured with a RotoVisco RV1 precision rotational viscometer (HAAKE).

Next, 0.1 part of platinum catalyst (Pt concentration, 1%) was mixed into this silicone rubber mixture to form a silicone rubber composition, and the composition was pre-cured at 120°C for 10 *min* and then post-cured in an oven at 150°C for 1 *h*. The resulting cured product was measured for hardness, tensile strength, elongation at break and tear strength.

Another addition curing silicone rubber composition has been described (11). In this composition, the amount of the low-molecular-weight siloxane component volatilizing from the cured product has been reduced, and as a consequence, this composition is free from problems such as clouding or haze, contact fault, adhesion failure, and hydrophobicization of the surface caused by the deposition of the volatilized low-molecular-weight siloxane component on the surrounding surfaces.

When a low-molecular-weight siloxane that does not contain a functional group volatilizes from the cured rubber and deposits on the surrounding surface, it can be readily removed, for example, by wiping with a solvent. In contrast, the removal of a low-molecular-weight cyclic siloxane component having a reactive group is not easy because of its reactivity, which results in a strong adhesion of the component to the substrate. A siloxane having a Si–H functional group is far more problematic than the low-molecular-weight siloxane having no functional group such as dimethyl siloxane (11).

2.1.6 *Liquid Injection Molding*

Liquid silicone rubber (LSR) or liquid injection molding compositions have typically involved two components being pumped in a ratio of 1:1 to 9:1 into a static mixer (12). This mixture can then be further processed by an injection molding machine or by other machines to produce moldings, extrudates, or coatings. The two components are composed minimally of poly(dimethyl siloxane)s having vinyl groups. The viscosities generally range from 0.65 to 10,000,000 $mPa\,s$. In order to achieve the final properties, fillers and additives are generally added.

Fillers can be active (reinforcing) or inert, depending on the requirements placed upon the LSR. Examples of active fillers are hydrophobic or hydrophilic highly dispersed silica, hydrophobic or hydrophilic precipitated silica, hydrophobic or hydrophilic aluminum oxide. Examples of inert fillers are powdered quartz, talc, and wollastonite.

A platinum catalyst such as hexachloroplatinic acid or a vinyl siloxane complex thereof is generally added to the first component. An Si–H compound which contains at least 2 or 3 silicon-bonded hydrogen atoms per molecule is added to the second component. An inhibitor is added into one of the components to regulate the crosslinking rate.

The components can also contain adhesion promoters or oils incompatible with the matrix, such as phenylsilicone oil, or heat stabilizers such as metal oxides, metal octoates, or carbon black compounds (12).

2.1.7 *Hot Embossing*

It has been shown that hot embossing is a viable technology for imprinting microstructures in addition to curing thin silicone films at their gel point (13). This is one of the simplest, most cost-effective, and time-saving methods for replicating microstructures.

Films made from LSR formulations containing fillers were hot embossed under modified operating conditions (13). The use of such relatively hard silicone elastomers shows the versatility of this method that has been established for softer silicone elastomers.

Also, as proof of this concept, a microstructured nickel plate could be successfully used as an embosser for the films. The ideal condition for hot embossing the LSR formulation (XLR 630 with titanium dioxide fillers) with a Ni embosser is 110°C preheating for 15 *s* to 35 *s*, embossed with 2 *bar* pressure, and postheating for complete curing at 110°C for 3 *min*. This shows that this process is extremely fast (13, 14). Also, a hot roller embossing process has been examined (15). These processes provide products with a strictly set network of microchannels (16).

2.2 Hydrosilylation Catalysts

The hydrosilylation of unsaturated compounds can be carried out by catalysis. Typically, the appropriate catalyst for this reaction is a platinum catalyst. Currently, the majority of industrial hydrosilylation reactions are catalyzed by the Karstedt platinum complex, with the general formula of Pt_2(divinyltetramethyldisiloxane)$_3$, c.f. Figure 2.2.

However, the use of platinum catalysts is still problematic because it is an expensive metal, which is becoming scarcer, and its cost fluctuates enormously. Its use on an industrial scale is difficult. It is thus desirable to reduce as much as possible the amount of catalyst necessary for the reaction, without, however, reducing the yield and the speed of the reaction.

Furthermore, it is desirable to have a catalyst available which is stable during the reaction. It has been found that, during the

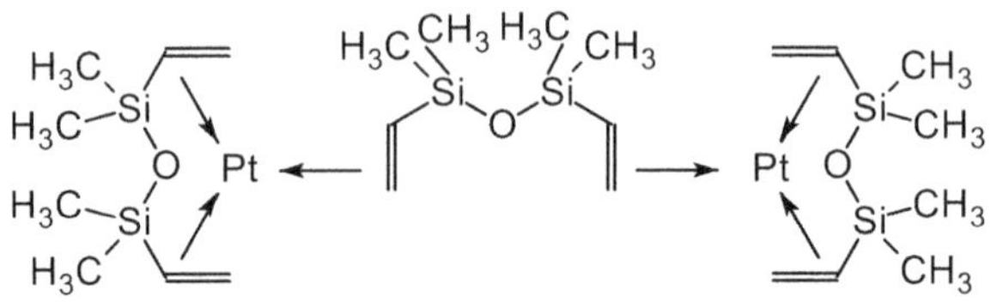

Figure 2.2 Karstedt platinum complex.

catalyzed reaction, platinum metal could precipitate, which has the consequence of forming colloids which are insoluble in the reaction medium. The catalyst is then less active. Furthermore, these colloids form a haze in the reaction medium and the products obtained are not esthetically satisfactory as they are colored.

In addition, during hydrosilylation reactions with the Karstedt catalyst, the formation of undesirable by-products has been observed, resulting from the isomerization of an olefinic double bond and/or from hydrogenation reactions. It is also desirable for the catalyst not to promote side reactions.

It has been suggested to use silylene ligands that are stabilized by Lewis bases (17). Some stable metal complexes coordinated by a silylene ligand stabilized by a Lewis base have been described in the literature (18). However, the complexes described are still not very stable and problems of stability are observed in these publications.

Metal complexes have been prepared where at least one of the ligands is a cyclic silylene compound stabilized by a Lewis base (17). The cyclic structure of these materials stabilized the silylenes. This makes it possible to obtain particularly stable metal complexes.

The methods of synthesis have been described in detail (17).

2.3 Recoating Methods

The effect of different recoating methods on the performance of room temperature vulcanized silicone rubber was studied (19).

Old and new coatings of room temperature vulcanized silicone rubber were tested using the row grid method, scanning electron microscopy and water boiling method. It is found that the old coating of room temperature vulcanized silicone rubber produces internal stress, as well as swelling in the recoating, and that surface-contaminating particles need to be removed before the recoating. Thus, the spraying method for recoating is recommended (19).

2.4 Shaped Elastomeric Bodies

Silicon elastomer articles can be prepared by a 3D printing process by depositing droplets of high viscosity curable silicone onto a substrate layer by layer from independently spatially controllable nozzles and irradiation with independently spatially controllable electromagnetic energy (20).

The apparatus for the production is shown in Figure 2.3.

The process for the layer-by-layer production of shaped silicone elastomer parts involves the following steps (20):

1. An independently spatially controllable jetting apparatus (1) is used for dropwise application (6) in the XY work plane, of at least one silicone rubber composition curable by electromagnetic radiation, via at least one printing head (5), to an independently spatially controllable baseplate (3) or to a shaped body affixed thereto.
2. At least one independently spatially controllable source of electromagnetic radiation (2) is used for crosslinking or incipient crosslinking (7) of the drop or drops (6) to form a layer of the cured or partially cured shaped silicone elastomer part.
3. The jetting apparatus (1) and/or the shaped silicone elastomer part from Step 2 are displaced relative to each other in the z-direction only far enough that the next drop layer (6) can be applied in the XY work plane.
4. Steps 1 to 3 are repeated until the construction of the shaped silicone elastomer part (8) is complete.

The silicone rubber composition used is addition crosslinking and is crosslinked in Step 2 by means of electromagnetic radiation (7) with induction either thermally and/or by UV or UV-VIS light. The silicone rubber composition has a viscosity of at least 10 $Pa\,s$, measured at 25°C, and a shear rate of 0.5 s^{-1}. The drops (6) in the XY work plane are exposed to the electromagnetic radiation

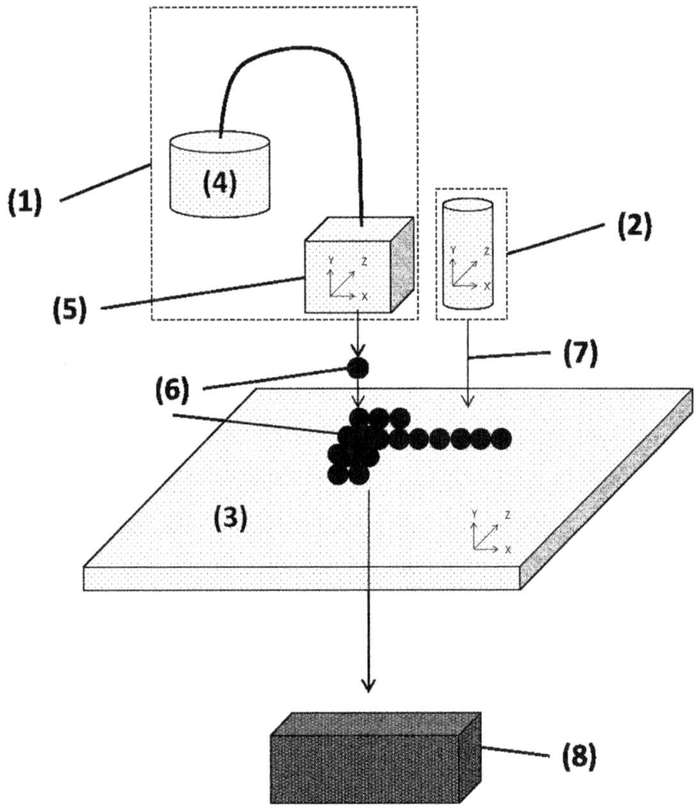

(1)	Jetting apparatus	(2)	Electromagnetic radiation
(3)	Controllable baseplate	(4)	Reservoir
(5)	Printing head	(6)	Drop layer
(7)	Crosslinking unit	(8)	Shaped elastomer part

Figure 2.3 Apparatus for 3D printing (20).

(7) site-selectively or areally, in pulses or continuously, and with constant or variable intensity (20).

The silicone rubber compositions that are used crosslink by a hydrosilylation reaction between Si-bonded hydrogen atoms and aliphatically unsaturated groups which are located preferably on the silicone polymer. Silicone rubber compositions of this kind are, for example, a RTV-2 (2-part room temperature vulcanizing) or a LSR.

Since the platinum-catalyzed addition reaction proceeds spontaneously even at room temperature, these silicone rubber compositions are typically supplied in the form of two components, with one component containing the SiH-functional crosslinking agent and the other component containing the platinum catalyst. However, by means of inhibitory additives, or through the use of specific platinum catalysts which are largely inert at room temperature but can be activated thermally, it is also possible to formulate one-component silicone rubber compositions, as described elsewhere (21).

The silicone rubber composition which can undergo addition-crosslinking under UV/VIS induction is cured with light of wavelength 250 *nm* to 350 *nm*. In order to achieve a rapid crosslinking reaction rate, by which is meant a crosslinking time at room temperature of less than 20 *min*, most preferably less than 1 *min*, it is advisable to use a UV/VIS radiation source having an output of preferably between 30 $mW\,cm^{-2}$ and 15,000 $mW\,cm^{-2}$, and also a radiation dose of preferably between 500 $mJ\,cm^{-2}$ and 10,000 $mJ\,cm^{-2}$. Within these output and dose values, area-specific irradiation times can be realized of between 2000 $s\,cm^{-2}$ at most and 8 $s\,cm^{-2}$ at least (20).

2.4.1 *Tailoring of Elastomers*

Specific properties related to soft applications have been reviewed concerning recipes and used to process elastomer networks, such as stretchable electronics, without compromising the mechanical integrity of the elastomer (22).

For research on flexible electronics and other emerging technologies, the most commonly utilized silicone elastomer formulation is Sylgard 184, which is easier to process than most other commercially available silicone elastomers, due to the fact that the premixes have a low viscosity. Furthermore, curing is robust and not as sensitive to poisoning as other silicone elastomer formulations. However, Sylgard 184 is not suitable for all fields of research that require flexible and stretchable silicones.

When much softer networks are needed, the Sylgard 184 premixes are either mixed in non-stoichiometric ratios, or they are blended with softer types of commercially available elastomers, which compromise the mechanical integrity of the elastomer.

The first part of a formulation process is to decide which elastic modulus is being targeted. It is complicated to formulate reliable silicone elastomers with elastic moduli below 0.6 *MPa*, since this is the apparent elastic modulus of well-entangled silicone polymers. The entanglements in crosslinked polymers behave like crosslinks with a more dynamic behavior, since they can slide in between each other and some of them can also disentangle.

Recently, so-called bottlebrush elastomers have been developed, due to effectively changing the molecular entanglement weight (23, 24). More details on entanglements and the characteristic properties of silicone polymers can be found in the literature (25).

Alternatively, elastomers can be processed in the presence of a large amount of solvent, in order to dilute locked entanglements and thereby proceed below 0.6 *MPa*. However, this process is not always favorable, since the solvent may swell other components in the devices (22).

If an optical transparency is desired, it is favorable to use resins rather than silica particles to reinforce the network, since there is a noticeable difference in the refractive index between silica and silicone polymers.

It is advisable to cure commercial silicone elastomers at temperatures provided by the manufacturer, although curing at other temperatures is possible. Different curing temperatures often result in elastomers with different mechanical properties (26,27).

Some recommended formulations are shown in Tables 2.2 and 2.3.

2.4.2 *Reinforcement of Elastomers*

A comprehensive review concerning the role of polyhedral oligomeric silsesquioxane (POSS) as reinforcing fillers for polymer

Table 2.2 Recommended formulations for hard silicone elastomers (22).

	Traditional	Optical transparency	Low viscosity
Network			
r	1.1–1.4	1.1–1.4	1.1–1.4
f	8	8	8
M $kg\ mol^{-1}$	20–70	20–70	< 20
Include	Silica	Resins	Resins Chain extender
Exclude	—	Silica	Silica

f = Functionality of crosslinking agent
M = Molecular weight of the silicone polymer in premixes before crosslinking reaction
r = Stoichiometry

Table 2.3 Recommended formulations for soft silicone elastomers (22).

	Adhesive	Moderately extensible	Highly extensible
Network			
r	<1	<1	<1
f		≥ 8	4–5
M $kg\ mol^{-1}$	≥ 70	≥ 70	≥ 70
Include	Resins for better track	Resins low loading	Silica low loading
Exclude	Silica	Silica	—

f = Functionality of crosslinking agent
M = Molecular weight of the silicone polymer in premixes before crosslinking reaction
r = Stoichiometry

and rubber nanocomposites has been presented (28). POSS, having an SiOSi framework, provides a high thermal stability, while the peripheral organic functionality offers the better polymer-filler interface adhesion.

The peripheral organic functionalities of POSS were reported to play a key role in controlling the POSS dispersion in fumed silica-based addition-cured liquid silicone rubber. The radical grafting of organo-functionalized POSS was reported to reinforce the ethylene propylene diene monomer (EPDM)-based rubbers, whilst causing a detrimental effect on radically curable polysiloxane elastomers.

On the other hand, the matrix-filler incompatibility and the preferential crystallization of POSS segments were effectively leveraged as tunable linkages to fabricate thermoplastic elastomeric materials. The use of POSS-based molecular units were also explored for developing functional structures like radiation protective coatings, nanoporous organic-inorganic frameworks for gas storage, or in low dielectric composites (28).

The incorporation of nanofillers into a polymeric matrix has been proven to be an effective route to reinforce their mechanical properties (29). The usage of binary fillers combines the advantages of the two fillers and could lead to a further enhancement of the properties.

Binary nanofillers consisting of multi-walled carbon nanotube (CNT)-bridged fumed silica were synthesized by a Pt coupling reaction of methyldiethoxysilane, c.f. Figure 2.4, modified multi-walled CNTs and triethoxyvinylsilane-functionalized SiO_2, in which methyldiethoxysilane modified multi-walled CNTs were synthesized by reacting OH-functionalized multi-walled CNTs with methyldiethoxysilane, while triethoxyvinylsilane-fumed silica was obtained by reacting fume silica with triethoxyvinylsilane (29).

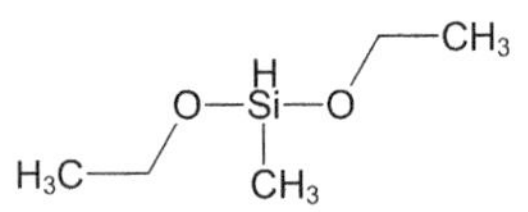

Figure 2.4 Methyldiethoxysilane.

The binary multi-walled CNTs-bridged fumed silica was introduced into LSR to achieve a mechanical reinforcement. It was shown that the interfacial interaction between binary fillers and LSR matrix was significantly enhanced due to the chemical bridging, as well as the excess triethoxyvinylsilane molecules on SiO_2. Here, the Young's modulus, tensile strength and the tear strength of the LSR could be increased by 64%, 29%, and 52%, respectively, with the incorporation of only 0.25 *phr* of multi-walled CNTs for the binary fillers.

The observed mechanical enhancement could be attributed to good interfacial interaction between the fillers and LSR due to the existence of SiO_2 with excess matrix compatible molecules on the surface of multi-walled CNTs, which facilitate the filler dispersion in LSR matrix and promote stress transfer from matrix to multi-walled CNTs, as well as the advantages of extremely high aspect ratio and excellent mechanical strength of multi-walled CNTs (29).

References

1. C.-C. Kuo and Z.-Y. You, *The International Journal of Advanced Manufacturing Technology*, Vol. 96, p. 1003, April 2018.
2. F. Liravi and M. Vlasea, *Additive Manufacturing*, Vol. 21, p. 112 , 2018.
3. B.S. Park, K.W. Seo, J.H. Lee, Y. Chun, S.H. Jin, B.K. Woo, and D.I.S. Luan, Silicone rubber composition and manufacturing method thereof, US Patent 9 963 577, assigned to Hyundai Mobis Co., Ltd. (Seoul, KR), Grace Continental Korea Co., Ltd. (Bucheon-si, KR), May 8, 2018.
4. N. Abdullah, M.S.A. Hossain, A. Fatehmulla, W.A. Farooq, M.T. Islam, N. Miyamoto, Y. Bando, Y. Kamachi, V. Malgras, Y. Yamauchi, and N. Suzuki, *Journal of Nanoscience and Nanotechnology*, Vol. 18, p. 86, 2018.
5. B.G. Trewyn, I.I. Slowing, S. Giri, H.-T. Chen, and V.S.-Y. Lin, *Accounts of Chemical Research*, Vol. 40, p. 846, 2007.
6. K. Ariga, Y. Yamauchi, G. Rydzek, Q. Ji, Y. Yonamine, K.C.-W. Wu, and J.P. Hill, *Chemistry Letters*, Vol. 43, p. 36, 2014.
7. V. Malgras, Q. Ji, Y. Kamachi, T. Mori, F.-K. Shieh, K.C.-W. Wu, K. Ariga, and Y. Yamauchi, *Bulletin of the Chemical Society of Japan*, Vol. 88, p. 1171, 2015.
8. Y. Li, B.P. Bastakoti, and Y. Yamauchi, *APL Materials*, Vol. 4, p. 040703, 2016.
9. M. Igarashi, M. Yoshida, Y. Mayumi, T. Kimura, and M. Ikeno, Silicone rubber composition, silicone rubber molded article, and production

method thereof, US Patent 8 952 077, assigned to Shin-Etsu Chemical Co., Ltd. (Chiyoda-ku, JP), February 10, 2015.
10. N. Kato and S. Shudo, Addition-curable liquid silicone rubber composition, US Patent 9 909 008, assigned to Shin-Etsu Chemical CO., LTD. (Tokyo, JP), March 6, 2018.
11. N. Meguriya, S. Ide, and T. Yamazaki, Addition curing silicone rubber composition and its cured product, US Patent 9 045 637, assigned to Shin-Etsu Chemical Co., Ltd. (Tokyo, JP), June 2, 2015.
12. P. Jerschow and E. Schaffer, Liquid silicone rubber, US Patent Application 20 080 125 528, assigned to Wacker Chemie AG, Munich (DE), May 29, 2008.
13. S. Vudayagiri, L. Yu, and A.L. Skov, *Journal of Elastomers & Plastics*, Vol. 47, p. 585, 2015.
14. M. Gaitan and L.E. Locascio, *Journal of Research of the National Institute of Standards and Technology*, Vol. 109, p. 335, June 2004.
15. C.-W. Tsao, T.-Y. Chen, W.Y. Woon, and C.-J. Lo, *Microsystem Technologies*, Vol. 18, p. 713, 2012.
16. N. Zolotareva and V. Semenov, *Silicon*, Vol. 7, p. 89, April 2015.
17. A. Baceiredo, T. Kato, R. Rodriguez, A. Prades, S. Marrot, and L. Saint-Jalmes, Catalysts with a silylene ligand, US Patent 9 938 304, assigned to Bluestar Silicones France SAS (Lyon, FR), Centre National de la Recherche Scientifique–CNRS (Paris, FR), April 10, 2018.
18. R. Rodriguez, T. Troadec, T. Kato, N. Saffon-Merceron, J.-M. Sotiropoulos, and A. Baceiredo, *Angewandte Chemie*, Vol. 124, p. 7270, 2012.
19. M. Zhu, Q. Zeng, and X. Gao, Effect of different recoating methods on the performance of room temperature vulcanized silicone rubber: An experimental study, in Q. Xu, ed., *Application of Materials Science and Environmental Materials (AMSEM2015)*, The 3rd Internatonal Conference on Application of Materials Science and Environmental Materials (AMSEM2015), pp. 75–81, New Jersey, 2015. World Scientific.
20. E. Selbertinger, F. Achenbach, and B. Pachaly, Method for producing silicone elastomer parts, US Patent Application 20 170 312 981, assigned to Wacker Chemie AG, Munich (DE), November 2, 2017.
21. A. Fehn and F. Achenbach, Curable organopolysiloxane materials, US Patent 6 359 098, assigned to Wacker-Chemie GmbH (Munich, DE), March 19, 2002.
22. P. Mazurek, S. Vudayagiri, and A.L. Skov, *Chemical Society Reviews*, Vol. 48, p. 1448, 2019.
23. L.-H. Cai, T.E. Kodger, R.E. Guerra, A.F. Pegoraro, M. Rubinstein, and D.A. Weitz, *Advanced Materials*, Vol. 27, p. 5132, 2015.
24. J. Paturej, S.S. Sheiko, S. Panyukov, and M. Rubinstein, *Science Advances*, Vol. 2, 2016.
25. L.J. Fetters, D.J. Lohse, D. Richter, T.A. Witten, and A. Zirkel, *Macromolecules*, Vol. 27, p. 4639, 1994.

26. F. Schneider, T. Fellner, J. Wilde, and U. Wallrabe, *Journal of Micromechanics and Microengineering*, Vol. 18, p. 065008, April 2008.
27. I.D. Johnston, D.K. McCluskey, C.K.L. Tan, and M.C. Tracey, *Journal of Micromechanics and Microengineering*, Vol. 24, p. 035017, February 2014.
28. D. Dasgupta, M. Srividhya, A. Sarkar, M. Dubey, D. Wrobel, and A. Saxena, Rubber nanocomposites with polyhedral oligomeric silsesquioxanes (POSS) as the nanofiller in S. Thomas and H.J. Maria, eds., *Progress in Rubber Nanocomposites*, Woodhead Publishing Series in Composites Science and Engineering, chapter 7, pp. 23–247. Woodhead Publishing, 2017.
29. J. Kong, J. Sun, Y. Tong, Q. Dou, Y. Wei, W. Thitsartarn, J.C.C. Yeo, and C. He, *Composites Science and Technology*, Vol. 169, p. 232 , 2019.

3

Automotive and Underwater Applications

3.1 Automotive Applications

3.1.1 Turbocharger Hose

Fluorosilicone rubber compositions, particularly those which are based on an organopolysiloxane polymer with a large proportion of perfluoroalkyl groups, when cured, possess properties rendering them resistant to heat, frost, chemicals, and exposure to oil (1). These compositions are therefore widely used in the automotive and aircraft industries.

However, one major problem with cured fluorosilicone rubber is that it does not adhere well to other substrates such as other silicone rubber compositions.

Various methods have been proposed for improving adhesion of fluorosilicone rubbers to alternative silicone rubber-based substrates and other substrate surfaces. These include the addition of materials containing multiple silicon-hydrogen bonds or multiple alkenyl groups in either the fluorosilicone rubber or other silicone rubber to enhance the adhesion. The introduction of adhesion promoters, such as cyanurate- and/or isocyanurate-based compounds, for example triallyl cyanurate or triallyl isocyanurate in either the fluorosilicone rubber or other silicone rubber, may also be used, but can result in compatibility problems. These compounds used here are shown in Figure 3.1.

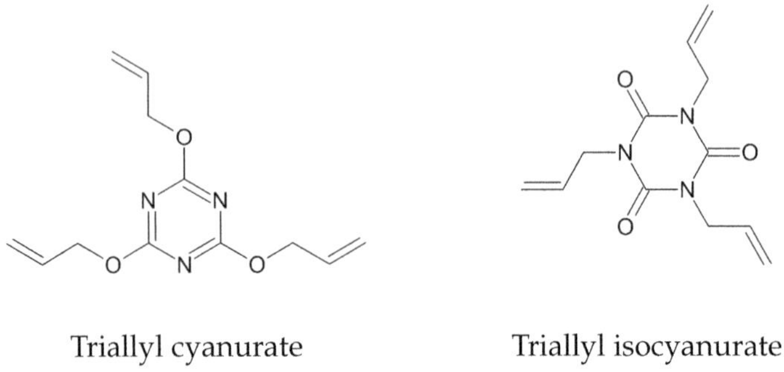

Figure 3.1 Cyanurate- and isocyanurate-based compounds.

A fluorosilicone rubber composition has been developed. The components are shown in Table 3.1.

Table 3.1 Fluorosilicone rubber composition (1).

Amount/parts per weight	Compound
67	Hydroxy-terminated trifluoropropylmethyl siloxane vinylmethyl siloxane copolymer
23	Fumed silica
8	Hydroxy-terminated trifluoropropylmethyl siloxane
0.3	Dimethylvinylsiloxy-terminated dimethyl, methylvinyl siloxane
1.8	Trimethylsilyl-group-terminated methyl(perfluorobutylethyl) siloxane methylhydrogen siloxane copolymer
1	Paste of 45% 2,5-Dimethyl-2,5-di(*tert*-butylperoxy)hexane catalyst in silicone oil

The preparation of samples was done as follows (1):

Preparation 3–1: Un-cured slabs having dimensions of 150×150×1 *mm* of fluorosilicone rubber, high consistency rubber, and mixture of fluorosilicone rubber and high consistency rubber were pressed out. These were then coupled together whilst inserting a PET sheet between the layers at one end to allow separation after cure. The coupled sheets were then cured in a 2 *mm* deep mold for 10 *min* at 170°C and post cured for 4 *h* at 200°C in a hot air oven. Then, 25 *mm* strips were cut using a suitable cutting tool and the resulting test samples were analyzed for the level of adhesion between the two layers.

Of the so-prepared samples, four were subjected to a 180° peel test using an Instron tensiometer at a crosshead speed of 50 $mm\ min^{-1}$.

The adhesion of mixture samples of high consistency rubber and fluorosilicone rubber in the mixing ratio from 60:40 to 40:60 gave significantly improved peel force test results with both high consistency rubber layer and fluorosilicone rubber layer (1).

The material can be used for fluorosilicone rubber compositions to be used in situations where improved oil resistance properties are required. One application which relates to both of these potential applications is in the preparation of parts for automobiles such as turbocharger hoses.

A typical turbocharger hose is depicted in Figure 3.2. The turbocharger hose (1) has an inner fluorosilicone liner (2), a mixture layer (3), a high consistency silicone rubber inner layer (4), a reinforcing layer (5), typically made from a suitable reinforcing fabric such as aramid or the like and a high consistency silicone rubber outer layer (6).

3.1.2 *Automotive Airbags*

The automobile airbag was invented by John W. Hetrick (2) and Walter Linderer (3), who filed their patents in the early 1950s.

Later research during the 1960s showed that compressed air could not inflate the mechanically based airbags fast enough for maximum safety, leading to the current chemical and electrically based airbags (4).

In 1967, a breakthrough occurred in the development of airbag crash sensors when Allen K. Breed (5) invented a mechanically based ball-in-tube component for crash detection, an electromechanical sensor with a steel ball attached to a tube by a magnet that would inflate an airbag in less than 30 *ms*.

In general, airbag devices are mounted on vehicles such as automobiles, where airbags are expanded and developed in cars in the event of an emergency, e.g., crash and rapid deceleration, so as to absorb shocks applied to occupants. Various types of airbag devices have been developed and adopted (6).

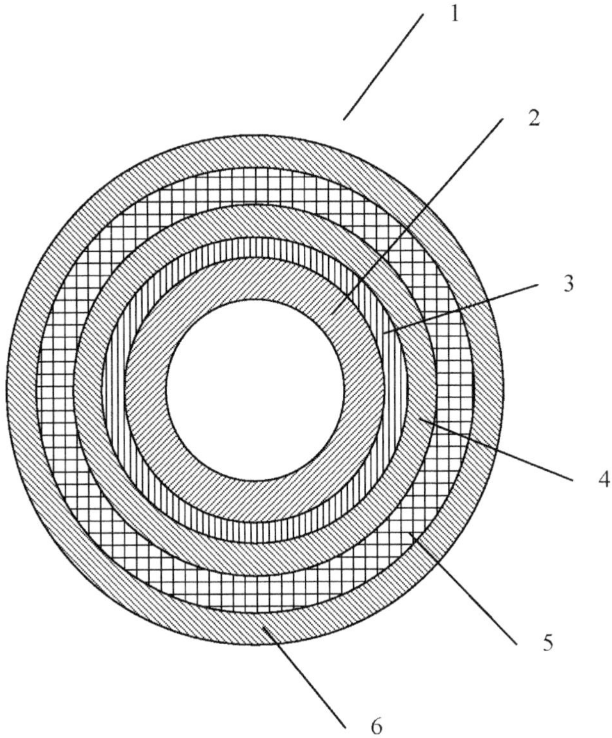

Figure 3.2 Turbocharger hose (1).

1 Turbocharger hose
2 Inner fluorosilicone liner
3 Mixture layer
4 High consistency silicone rubber inner layer
5 Reinforcing layer
6 High consistency silicone rubber outer layer

Examples of airbag devices include a driver airbag device incorporated in a steering wheel, a passenger airbag device incorporated in an instrument panel, a side airbag device incorporated in a vehicle side portion or a seat, a curtain airbag device incorporated in an upper portion of a door, and a knee airbag device corresponding to an an occupant's knees (6).

3.1.2.1 *Airbag Inflator*

A silicone rubber sheet has been developed that is suitable for use as a flow passage component of a high-temperature high-pressure gas spouted from an inflator of an airbag device mounted on a vehicle including a silicone rubber sheet (6).

In general, the gas spouted from the inflator is a high-temperature high-pressure gas and the vicinity of the spouting nozzle of the inflator has a heat-resistant pressure-resistant structure. In particular, it is necessary that the curtain airbag, i.e., the head side airbag can be expanded and developed in a wide range from a front seat to a rear seat and, in many cases, a large high-power inflator is employed to feed gas into the airbag.

A silicone rubber sheet is used for a flow passage to feed gas from an inflator, which generates gas by burning an agent stored in the inside, to an airbag that is usually folded and is expanded and developed in the event of an emergency, and the silicone rubber sheet is produced by stacking 0.5 *mm* or more of the silicone composition.

The silicone compositions consist of (6):

1. Organopolysiloxane,
2. Reinforcing silica,
3. Curing agent, and
4. Adhesion aid.

The organopolysiloxane component (1) has far less strength than that of other synthetic rubbers and is not at a level of being usable by itself. Therefore, a filler, in particular reinforcing silica, is added to the organopolysiloxane and, thereby, the strength suitable for use can be exhibited.

Examples of the reinforcing silica include fumed silica (dry silica) and precipitated silica (wet silica). In particular, fumed silica (dry silica) with a small impurity content is preferable. Also, the surface of the reinforcing silica may be hydrophobized with organopolysiloxane, organosilazane, chlorosilane, alkoxysilane, etc.

As curing agents for the silicone composition, organic peroxides, such as di-*tert*-butyl peroxide, 2,5-dimethyl-2,5-di(*tert*-butylperoxy)hexane, and dicumyl peroxide can used for a radical reaction. For addition reaction curing, agents can be used that are composed of organohydrogenpolysiloxanes having at least two hydrogen atoms bonded to silicon atoms in one molecule and platinum group metal-based catalysts in the case where the organopolysiloxane has an alkenyl group, and organosilicon compounds serving as condensation reaction curing agents and having at least two hydrolyzable groups, such as an alkoxy group, an acetoxy group, a ketoxime group, and a propenoxy group, in the case where the organopolysiloxane contains a silanol group. The amount of addition thereof may be the same as the amount in a common silicone rubber.

Examples of adhesion aids can include organosilicon-based compounds having preferably two or more types of functional groups selected from the group consisting of an alkenyl group, e.g., a vinyl group, a (meth)acryloxy group, a hydrosilyl group, an epoxy group, an alkoxysilyl group, a carbonyl group, a phenyl group, and an amino group and also non-silicon-based organic compounds. Some examples are collected in Table 3.2 and in Figure 3.3.

Table 3.2 Adhesion aids.

Compound
Vinyltrimethoxysilane
Benzyltrimethoxysilane
Propyl acrylate trimethoxysilane
Propyl methacrylate trimethoxysilane
(3-Glycidyloxypropyl)trimethoxysilane
3-Aminopropyltrimethoxysilane

Furthermore, if necessary, various additives, such as a color pigment, a heat resistance improver, a flame retardancy improver, and an acid acceptor, and various types of alkoxysilane, diphenylsilane diol, carbon functional silane, silanol-containing siloxane, serving as fluorine-based mold release agents or reinforcing silica dispersing agents may be added.

The preparation of such a composition can be done as follows (6):

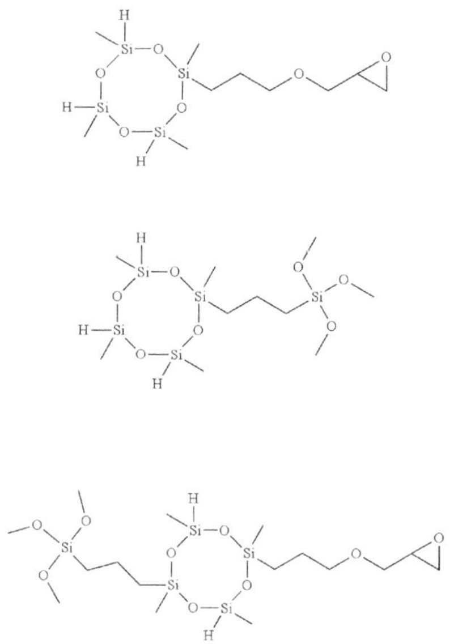

Figure 3.3 Adhesion aids (6).

Preparation 3–2: A base compound was obtained by kneading 100 parts by mass of methylvinylpolysiloxane containing 99.825 *mol%* of dimethyl siloxane unit, 0.15 *mol%* of methylvinyl siloxane unit, and 0.025 *mol%* of dimethylvinyl siloxane unit and having an average degree of polymerization of 8,000, 30 parts by mass of silica fine powder having a specific surface area of 200 m^2g^{-1} (Aerosil™ 200), 3 parts by mass of diphenylsilane diol, and 4 parts by mass of silanol-terminated dimethylpolysiloxane (degree of polymerization n=10) homogeneously and performing a heat treatment at 150°C for 2 *h*.

3.1.2.2 *Coated Airbag*

Airbags are generally formed from a woven or knitted fabric made of synthetic fiber, for example, a poly(amide) (PA) such as nylon-6,6 or a polyester, covered on at least one of its sides with a layer of elastomer (7). Airbags may be made of flat fabric pieces which are coated and then sewn together to provide sufficient mechanical strength, or may be woven in one piece with integrally woven seams. Sewn airbags are generally assembled with the coated fabric surface on the inside of the airbag.

One-piece woven airbags are coated on the outside of the airbag. A preferred elastomer for coating the airbag or airbag fabric is a silicone elastomer which is a cured organopolysiloxane composition, particularly a silicone rubber coating cured by hydrosilylation, by means of the reaction of alkenyl groups of one poly(organo siloxane) and Si–H groups of another poly(organo siloxane).

A process for coating an airbag or airbag fabric has been detailed, in which the bag or fabric is coated with a poly(organo siloxane) composition, characterized in that the airbag is top coated with an antifriction coating composition comprising at least one dispersed solid lubricant.

Also, an antifriction coating composition has been described that contains an aqueous dispersion of a solid fluoropolymer in a dispersion of an organic polymer binder, characterized in that the aqueous dispersion contains a siloxane poly(ether) wetting agent (7).

It has been found that a combination of a fluoropolymer and talc is particularly effective at reducing friction at the airbag surfaces. The addition of talc to an antifriction coating composition containing poly(tetrafluoroethylene) (PTFE) as solid lubricant can substantially reduce the coefficient of friction of the coated fabric, for example, by half or less of the coefficient of friction that can be achieved by PTFE alone, and much less than can be achieved by talc alone as solid lubricant.

The fluoropolymer and/or talc is in a dispersion of an organic polymer binder or an organic polymer binder/aminosilane mixture.

The mixture can be an aqueous solution, optionally containing a water-miscible organic cosolvent if required, or can be an aqueous emulsion or suspension. Such an emulsion or suspension is generally stabilized by at least one surfactant.

An example of a suitable non-ionic surfactant is an alkyl ethoxylate (ethoxylated fatty alcohol) or an aralkyl ethoxylate, such as (t-Octylphenoxy)polyethoxyethanol. An example of a suitable anionic surfactant is sodium dodecylbenzenesulfonate.

The selection of a suitable cosolvent is dependent on the organic polymer binder in the composition. Potential cosolvents include alcohols and *n*-alkylpyrrolidones.

Suitable amino silanes are collected in Table 3.3 and some are shown in Figure 3.4.

The preparation of an antifriction coating composition can be done as follows (7):

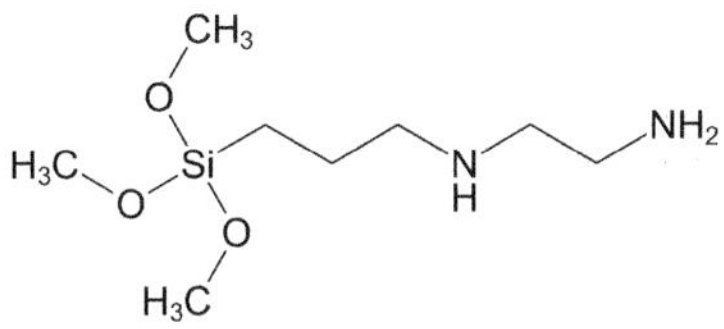

(Ethylenediaminepropyl)-
trimethoxysilane

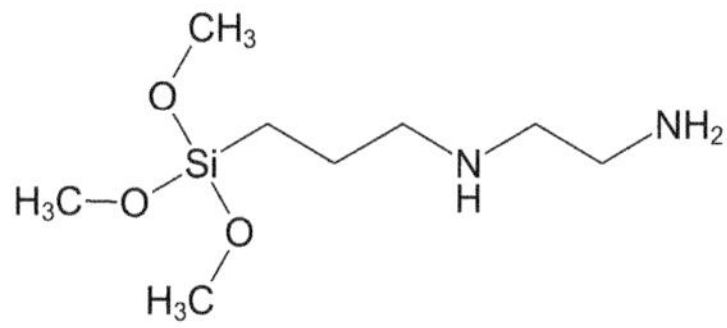

N-(2-Aminoethyl)-3-aminopropyl-
trimethoxysilane

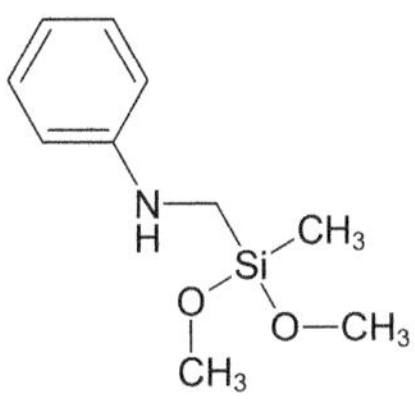

N-Phenylaminomethyl-
dimethoxymethylsilane

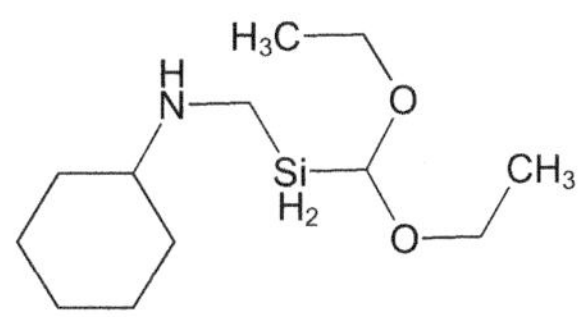

N-Cyclohexyl-aminomethyl-
dimethoxymethylsilane

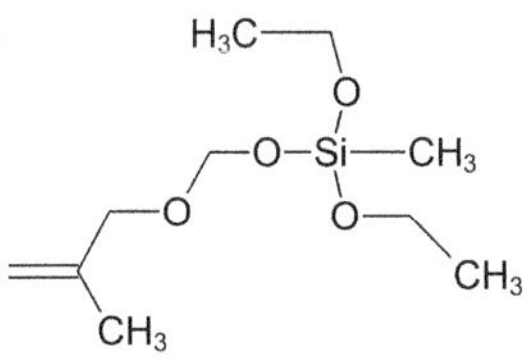

(Methacryloyl-oxymethyl)-
dimethoxymethylsilane

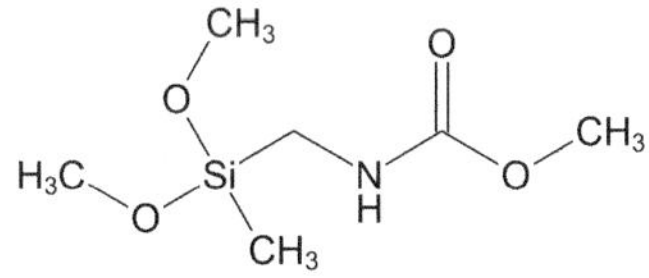

N-(Dimethoxymethyl-silylmethyl)-
o-methylcarbamate

Figure 3.4 Amino silanes.

Table 3.3 Amino silanes (7).

Compound
Ethylenediaminepropyltrimethoxysilane
N-(2-Aminoethyl)-3-aminopropyltrimethoxysilane
N-Phenylaminomethyldimethoxymethylsilane
N-Cyclohexyl-aminomethyldimethoxymethylsilane
N-Methylaminomethyldimethoxymethylsilane
N-Ethylaminomethyldimethoxymethylsilane
N-Propylaminomethyldimethoxymethylsilane
N-Butylaminomethyldimethoxymethylsilane
Methacryloyl-oxymethyldimethoxymethylsilane
N-(Dimethoxymethyl-silylmethyl)-*o*-methylcarbamate

Preparation 3–3: 1,1,1,3,5,5,5-Heptamethyl-3-polyethoxypropyl-trisiloxane wetting agent and 3-glycidoxypropyl trimethoxysilane adhesion promoter are added to an aqueous aliphatic poly(urethane) (PU) dispersion stabilized by a non-ionic surfactant, and mixed with an aqueous dispersion of PTFE powder stabilized by non-ionic surfactant to form an antifriction coating composition consisting of 52% water, 38% PTFE, 3.0% PU, 3.0% wetting agent, 1.0% epoxysilane adhesion promoter and 3.0% (*tert*-octylphenoxy)polyethoxyethanol non-ionic surfactant.

The antifriction coating was applied by gravure roller coating to the coated surface of a woven nylon airbag fabric coated with 55 $g\,m^{-2}$ of a cured liquid silicone rubber.

3.1.2.3 Coated Woven Fabric

When the coating amount of a curable silicone rubber composition applied to a woven fabric is reduced for the purpose of reduction of the weight of a curtain airbag, using a conventional known curable silicone rubber composition for coating woven fabrics, there is a problem in that the inflation duration of the airbag is excessively shortened (8).

A liquid curable silicone rubber composition has been developed for coating a woven fabric. The liquid curable silicone rubber composition is capable of sustaining a long inflation time for the inflated airbag, even when the coating amount of the composition applied to a woven fabric is reduced for the purpose of weight and cost reduction of an airbag.

Furthermore, a cured product of the liquid curable silicone rubber composition is excellent in terms of heat resistance and moisture resistance and is particularly appropriate as a curable silicone rubber composition for coating a woven fabric used for producing curtain airbags.

The liquid curable silicone rubber composition contains the following components (8):

1. An organopolysiloxane having one or more silicon atom--bonded alkenyl groups on average in one molecule,
2. An organohydrogenpolysiloxane having at least three silicon atom-bonded hydrogen atoms in one molecule,
3. A hydrosilylation reaction catalyst,
4. A curing retardant,
5. A reinforcing silica fine powder, and
6. An adhesion promoter.

The curing retardant is selected from acetylene compounds, enyne compounds, triazoles, and hydrazines. Specific examples are shown in Table 3.4 and in Figure 3.5.

Table 3.4 Curing retardants (8).

Compound	Compound
1-Ethynyl-1-cyclohexanol	2-Methyl-3-butyn-2-ol
3,5-Dimethyl-1-hexyn-3-ol	3,5-Dimethyl-1-octyn-3-ol
2-Phenyl-3-butyn-2-ol	3-Methyl-3-penten-1-yne
3,5-Dimethyl-3-hexen-1-yne	Benzotriazole

As adhesion promoters, most preferable compositions are a combination of the methacryloxy-containing organoalkoxysilane and the zirconium-chelate compound or a combination of the methacryloxy-containing organoalkoxysilane compound, the epoxy-containing organoalkoxysilane compound, and the zirconium-chelate compound (9). Specific examples are 3-glycidoxypropyl trimethoxysilane, 3-methacryloxypropyl trimethoxysilane, c.f. Figure 3.6, and a mixture of 50 parts by weight of a zirconium tetracetylacetonate with 50 parts by mass of a dimethylpolysiloxane that has a viscosity of 400 $mPas$ and is capped at both molecular terminals with dimethylvinylsiloxy groups.

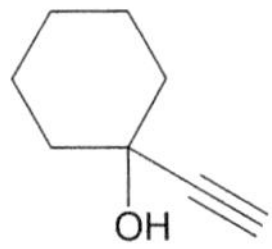

1-Ethynyl-1-cyclohexanol

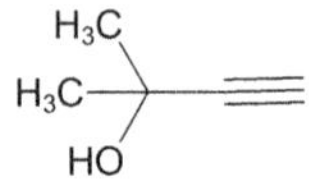

2-Methyl-3-butyn-2-ol

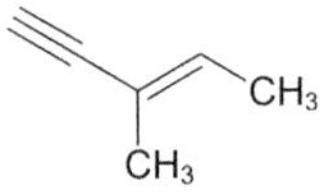

3-Methyl-3-penten-1-yne

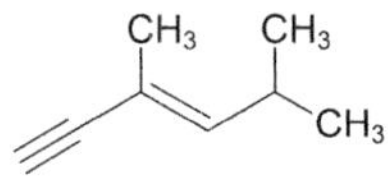

3,5-Dimethyl-3-hexen-1-yne

Figure 3.5 Curing retardants.

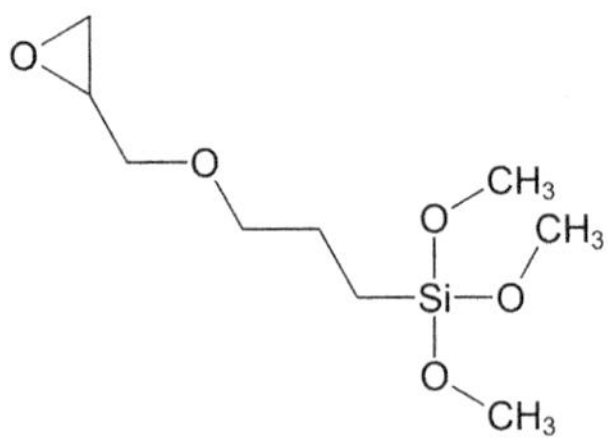

3-Glycidoxypropyl
trimethoxysilane

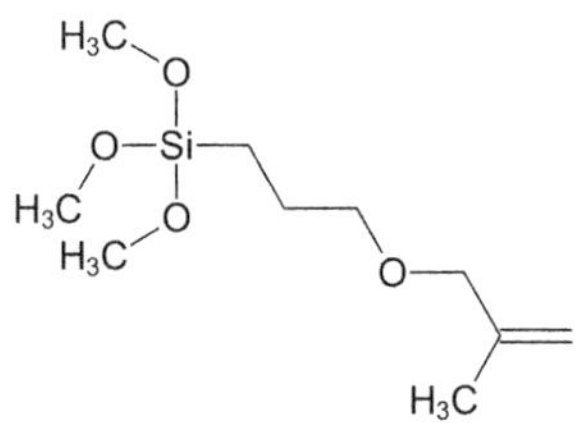

3-Methacryloxypropyl
trimethoxysilane

Figure 3.6 Adhesion promoters (9).

As curing inhibitor, a mixture of 2 parts by mass of an ethynylcyclohexanol, c.f. Figure 2.1, with 98 parts by mass of a dimethylpolysiloxane that has a viscosity of 10,000 $mPa\,s$ and is capped at both molecular terminals with dimethylvinylsiloxy groups can be used (9). Another curing inhibitor is 3,5-dimethyl-1-octyn-3-ol, c.f. Figure 3.7.

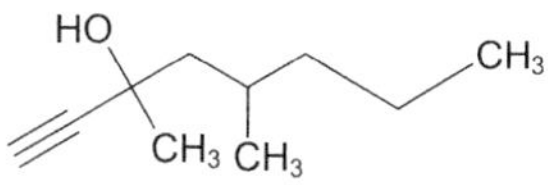

Figure 3.7 3,5-Dimethyl-1-octyn-3-ol.

In detail, a method of preparation of a flowable silica master batch can be performed as follows (8,9):

Preparation 3–4: A Ross mixer is charged with 100 parts by weight of a dimethyl(polysiloxane) that was terminated at both terminals of the molecular chain with dimethylvinylsiloxy groups and had a viscosity of 40,000 $mPa\,s$, 36 parts by weight of fumed silica having a BET specific surface area of 225 $m^2\,g^{-1}$, 7 parts by weight of a hexamethyldisilazane and 1.4 parts by weight of water. After the components were mixed to uniformity at room temperature, the obtained mixture was heat-treated for 2 h at 200°C under a reduced pressure, whereby a flowable silica master batch was prepared.

A cured silicone rubber-coated layer can be formed on a woven fabric by coating the liquid curable silicone rubber composition onto a woven fabric and then curing the composition (8). Examples of the woven fabric include fabrics made of PA fibers such as Nylon 6, Nylon 66 and Nylon 46, fabrics made of polyester fibers such as poly(ethylene terephthalate) and poly(butylene terephthalate), poly(acrylonitrile) fiber fabric, aramid fiber fabric, polyether imide fiber fabric, polysulfone fiber fabric, carbon fiber fabric, rayon fiber fabric, and poly(ethylene) fiber fabric, or a nonwoven fabric made from the aforementioned fibers. It is preferable to use PA fiber fabric or polyester fiber fabric for applications requiring high strength, such as automotive airbags.

Prior to coating with the liquid curable silicone rubber composition, the woven fabric is preferably washed with water and dried (8).

As a method for applying the liquid curable silicone rubber composition onto the woven fabric, known processes such as spraying, gravure coating, bar coating, knife coating, patting, screen printing, and dipping can be used. The coating amount of the curable composition on a woven fabric is typically in the range of 25 g/m^{-2} to 150 g/m^{-2}. Curing is typically carried out by heating the composition at a temperature of from 150°C to 200°C for 1–2 *min* (8).

The silicon rubber composition developed herein is suitable for use as a fabric coating composition since it allows the fabric to maintain a pressure barrier between two areas with a pressure differential over a relatively long time (9). Therefore, the composition can be used for coating textile fabrics utilized in the structure of vehicle side curtain airbags, aircraft emergency exit seats, and inflatable rafts.

When an organophosphazene compound, c.f. Figure 3.8, is used in a silicone rubber coating composition for airbag base fabric at a predetermined amount, the resulting liquid silicone rubber coating composition is capable of providing a coated fabric (10). Namely, the base fabric having a coating formed by curing the composition has an excellent low burning speed defined by FMVSS-302 (Federal Motor Vehicle Safety Standard) (11), compliant to ISO 3795 (12), and the cured coating layer also has a low surface tackiness with a high anti-blocking property (10).

3.1.2.4 *Fluorosilicone Rubber Composition*

Compositions based on fluorosilicone rubber having 3,3,3-trifluoropropyl groups as side chain have a solvent resistance as well and find use as diaphragms, O-rings, oil seals and other parts in transportation equipment and petroleum-related equipment (13).

However, it is known that fluorosilicone rubber substantially swells and loses its physical strength in alcohol-mixed fuels, which are increasingly used in modern vehicles. The rubber is found to experience a substantial loss of strength when immersed in FAM B, a test fuel oil consisting of toluene, isooctane, diisobutylene, methanol, ethanol and water. Efforts to develop fluorosilicones having enhanced durability to alcohol fuel have been made. However, these methods fail to improve the tensile strength of fluorosilicone after immersion in an alcohol-containing fuel oil.

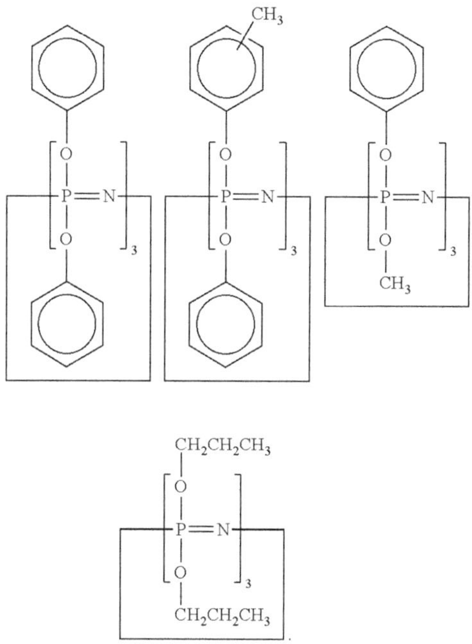

Figure 3.8 Organophosphazenes (10).

It has been found that a combination of two types of fluorosilicone gums with a reinforcing filler is effective for suppressing the degradation of physical strength of its cured product after immersion in alcohol-containing fuel oil (13). These fluorosilicone gums are shown in Figure 3.9.

To 100 parts by weight of each base compound, 1 part by weight of cerium oxide having a specific surface area of 130 m^2g^{-1} and 0.8 part by weight of a paste containing 80% of 2,5-dimethyl-2,5-di(*tert*-butyl-peroxy)hexane, c.f. Figure 3.10, were added. They were mixed on a twin-roll mill until uniform.

The composition was cured by press molding at 165°C for 10 *min*, and post-cured at 200°C for 4 *h*. Sheet specimens were prepared for 2 *mm* thick hardness measurement and dumbbell-shape specimens, as defined in JIS K-6249 (14), for tensile strength and elongation at break measurement. After immersion in a FAM B solvent mixture, the change of tensile strength was −40% and the change of elongation at break was −25% (13).

It was found that a loss of physical properties of a fluorosilicone

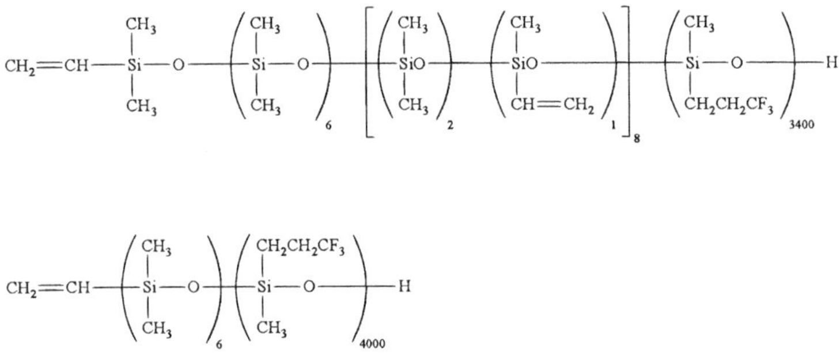

Figure 3.9 Fluorosilicone gums (13).

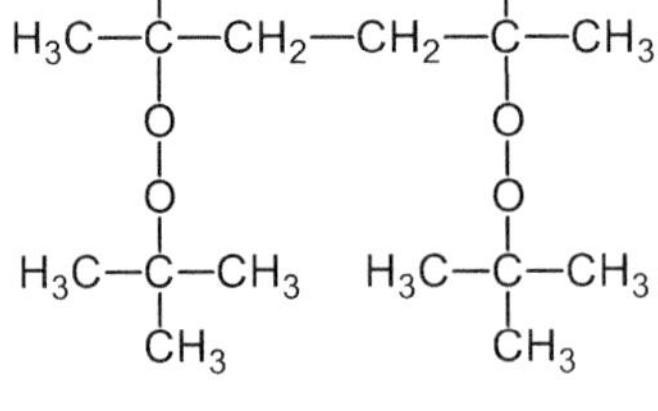

Figure 3.10 2,5-Dimethyl-2,5-di(*tert*-butyl-peroxy)hexane.

rubber can occur when brought in contact with an amine anti-aging agent (15). This can be suppressed by adding a particular type of activated carbon to the fluorosilicone rubber composition.

Known activated carbons include those derived from coal, wood, and coconut husk. The preferred type is those derived from wood and coconut husk in view of the well-developed fine pore structure. The activated carbon needs to have a pH of 6.5 to 8.0. It should have an average particle size of up to 70 μm. Recommended activated carbons are Shirasaqi AS (derived from wood) and FP-3 (derived from corn husk), which are products of Japan EnviroChemicals Ltd.

The activated carbon suppresses a cracking of the siloxane chain by the amine anti-aging agent. The cracking of the siloxane chain by the amine is suppressed by the adsorption by the minute pores and fixing by the surface of the carbonyl group (15).

The amine durability was evaluated by measuring the compression set after immersing in an amine. The measurement was conducted according to JIS K 6262 (16). After curing the resulting full compound by pressing under pressure at 165°C for 10 *min*, a post curing at 200°C was conducted for 4 *h* to prepare a small columnar test piece having a diameter of 13 *mm* and a height of 6.3 *mm* for use in the measurement of the compression set. The resulting test piece was immersed in a 10% solution of *N,N'*-di-*sec*-butyl-*p*-phenylenediamine, c.f. Figure 3.11, solution in ethanol for 24 *h*, and after air-drying at room temperature for 1 *d*, the 25% compression set was measured at 150°C (15).

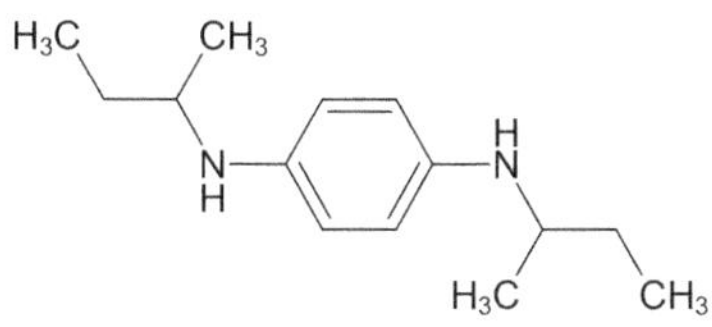

Figure 3.11 *N,N'*-di-*sec*-Butyl-*p*-phenylenediamine.

3.1.2.5 Liquid Silicone Rubber Coating Composition

Base fabrics comprising a synthetic fiber fabric, such as Nylon 66 coated with silicone rubber, offer the advantages of good thermal characteristics, good nonflammability, and excellent resistance to deterioration with the passage of time. These advantages have led to their use in automotive airbags (17, 18).

A diluting solvent, such as toluene or xylene, is generally used when coating the silicone rubber on the synthetic fabric in order to obtain a uniform thin coating and to achieve a strong bonding with the base fabric. However, a liquid silicone rubber composition that is coatable without the use of solvent has been introduced in response to the environmentally inspired drive to eliminate or reduce the use of solvents. A drawback to fabrics coated with this composition is that the surface of the applied film presents some residual tack even after cure into the silicone rubber. When left as is, this residual tack impairs the processability during such operations as sewing and also causes adjacent films of the coating to stick to each other during storage of the folded fabric. These problems have made it necessary to dust the surface of the applied films with, for example, talc, calcium carbonate, or clay.

A curable liquid silicone rubber coating composition has been developed (19). This composition exhibits excellent infiltrability and can be used for thin coatings on synthetic fabrics, without using a dilution solvent. The cured composition is a tack-free coating film with excellent adherence to synthetic fabrics. The curable composition contains (19):

1. A diorganopolysiloxane that contains at least 2 alkenyl groups in each molecule,
2. An organopolysiloxane resin,
3. An inorganic filler,
4. An organohydrogenpolysiloxane that contains at least 2 silicon-bonded hydrogen atoms in each molecule,
5. A platinum group catalyst,
6. An epoxy-functional organosilicon compound, and
7. An organotitanium compound.

Examples of diorganopolysiloxanes are dimethylvinylsiloxy-endblocked dimethylpolysiloxanes, dimethylvinylsiloxy-endblocked dimethyl siloxane-methylvinyl siloxane copolymers, dimethylvinylsiloxy-endblocked dimethyl siloxane-methylphenyl siloxane copolymers, and dimethylvinylsiloxy-endblocked methyl(3,3,3-trifluoropropyl)siloxane methylvinyl siloxane copolymers.

Inorganic fillers are fumed silica, precipitated silica and calcined silica, fumed titanium oxide, crushed quartz, diatomaceous earth, iron oxide, aluminum oxide, calcium carbonate, and magnesium carbonate. Here, ultrafine silica powders with specific surface areas of at least 50 m^2g^{-1} are preferred. The optimal filler is surface-treated ultrafine silica powder, for example, silica whose surface has been preliminarily treated with compounds such as organosilane, organosilazane and diorganocyclopolysiloxane.

The organohydrogenpolysiloxane acts as crosslinking agent for the composition. It contains at least two silicon-bonded hydrogen atoms in each molecule. This organopolysiloxane is exemplified by trimethylsiloxy-endblocked methylhydrogenpolysiloxanes, trimethylsiloxy-endblocked dimethyl siloxane-methylhydrogensiloxane copolymers, dimethylphenylsiloxy-endblocked methylphenyl siloxane-methylhydrogensiloxane copolymers, cyclic methylhydrogenpolysiloxanes, and copolymers composed of the dimethylhydrogensiloxy and $SiO_4/2$ units.

The platinum group catalyst used here is a curing catalyst for the composition. This component is exemplified by very finely divided platinum powder, platinum black, chloroplatinic acid, platinum tetrachloride, chloroplatinic acid-olefin complexes, alcohol solutions of chloroplatinic acid, chloroplatinic acid-alkenyl siloxane complexes, rhodium compounds, and palladium compounds.

The epoxy-functional organosilicon compound is an essential component that improves the adherence of the composition for the synthetic fabrics used in airbags. This compound is exemplified by epoxy-functional organoalkoxysilanes, such as γ-glycidoxypropyl trimethoxysilane, β-(3,4-epoxycyclohexyl)ethyltrimethoxysilane, and by epoxy-functional organopolysiloxanes, such as epoxy-functional organopolysiloxanes that also contain silicon-bonded vinyl and alkoxy, epoxy-functional organopolysiloxanes that also contain silicon-bonded hydrogen, epoxy-functional organopolysiloxanes that also contain silicon-bonded hydrogen and alkoxy.

The organotitanium compound is an essential component that improves the adherence of the composition and that reduces the post-cure tack of the surface of the film coating. This organotitanium compound is exemplified by organotitanate esters such as tetraisopropyl titanate, tetrabutyl titanate, tetraoctyl titanate, and by titanium chelate compounds such as diisopropoxybis(acetyl acetonato)titanium, diisopropoxybis(ethyl acetoacetate)titanium.

This component is added at from 0.1 to 5 weight parts per 100 weight parts component (A). The use of more than 5 weight parts of this component causes a deterioration in the storage stability of the composition. When used at less than 0.1 weight part, this component will not manifest its function of reducing the tack of the surface of the film coating after cure into the silicone rubber.

To produce a silicone rubber-coated fabric, the composition as described above is coated onto the synthetic fabric followed by introduction into a hot-air drying oven for heating and curing (19). The silicone rubber composition is generally applied at a rate not exceeding 80 $g\,m^{-2}$. The heating and curing step affords a flexible coated fabric in which the silicone rubber coating film and synthetic fabric have been tightly bonded into a single body at an adhesive strength of at least 1.5 $kgf\,cm^{-1}$.

Moreover, no tack is exhibited by the surface of the coating film on coated fabric prepared by application of the composition. As a result, even in the absence of dusting with, for example, talc or calcium carbonate, this tack-free condition for the subject coated fabric makes possible an excellent processability in operations such as sewing, and also eliminates sticking between adjacent coating films during storage in a folded state (19).

3.1.2.6 Curtain Airbag Device

An airbag device is shown in Figure 3.12. The airbag which is

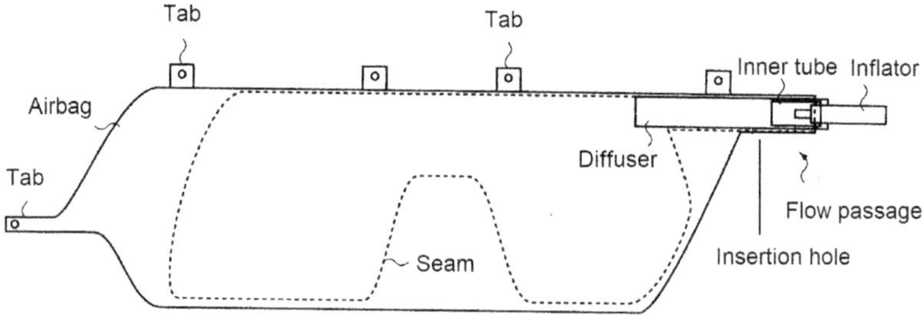

Figure 3.12 Airbag device (20).

usually folded and which is expanded and developed in the event of an emergency, contains an inflator, which generates gas by burning an agent stored in the inside, and a flow passage to feed the gas from the inflator to the airbag, wherein a silicone rubber sheet is used for at least part of the flow passage. The airbag is a so-called curtain airbag device, where the airbag is expanded and developed over a front seat and a rear seat.

For example, a bag body of the airbag is formed by bonding a base cloth to constitute a front surface together with a base cloth to constitute a back surface and sewing up the outer circumferences, and then the airbag is divided into a plurality of expansion bodies by a seam sewed in such a way as to form the expansion bodies, which are expanded into desired shapes. A plurality of tabs to be connected to a car body (for example, roof side rail) are disposed on the outer circumference of the airbag. The inflator of a so-called cylinder type explosive system is used. The inflator can be connected to an electronic control unit of a vehicle and is controlled on the basis of a measurement value of an acceleration sensor. When the electronic control unit detects or predicts crash or rapid deceleration of the vehicle, the inflator is ignited because of an ignition current from the electronic control unit, an agent stored in the inside of the inflator is burnt to generate gas, the gas is spouted from a spouting nozzle disposed at the end and, thereby, the gas is fed to the airbag.

The silicone rubber sheet described above is produced by stacking 0.5 *mm* or more of a silicone composition containing (20):

1. 100 parts by weight of an organopolysiloxane which is represented by the following average composition formula $R{-}SiO_{(4-n)/2}$, wherein the formula, R, is a substituted or unsubstituted monovalent hydrocarbon group and n represents a positive number of 1.95 to 2.05. The rubber has an average degree of polymerization of 1,000 or more,
2. 20 to 80 parts by weight of a reinforcing silica powder having a specific surface area of 50 m^2g^{-1} or more, and
3. A curing agent on at least an inner surface of a synthetic fiber fabric or an inorganic fiber fabric constituting flow passage and heat cure is performed thereon.

In addition, for the purpose of improving the adhesion, an adhesion aid may be blended. Examples of the adhesion aids can include organosilicon-based compounds having at least one type of functional groups selected from the group consisting of a vinyl group, a (meth)acryloxy group, a hydrosilyl group, an epoxy group, an alkoxysilyl group, a carbonyl group, a phenyl group, and an amino group and also non-silicon-based organic compounds. Specific examples of the adhesion aids may include the compounds shown in Figure 3.13.

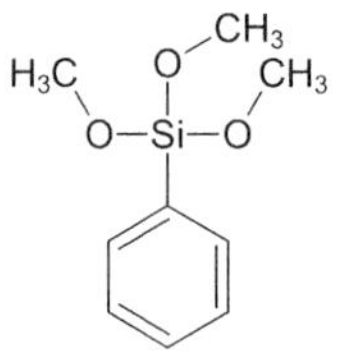

Phenyl trimethoxysilane

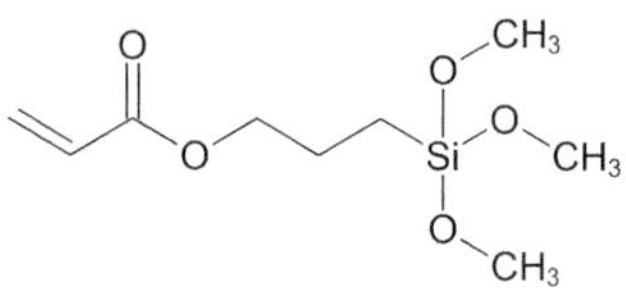

Acryloxy propyl trimethoxysilane

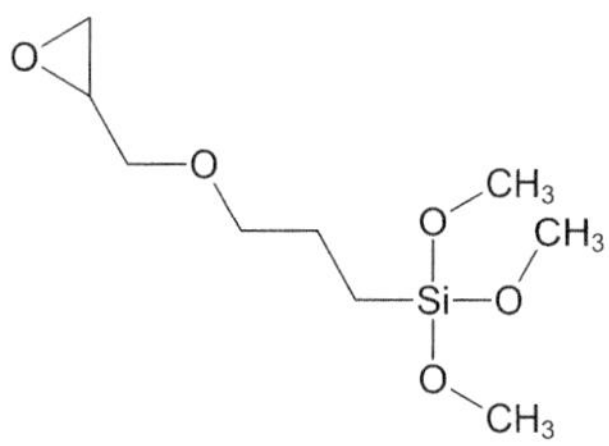

(3-Glycidyloxypropyl)trimethoxysilane

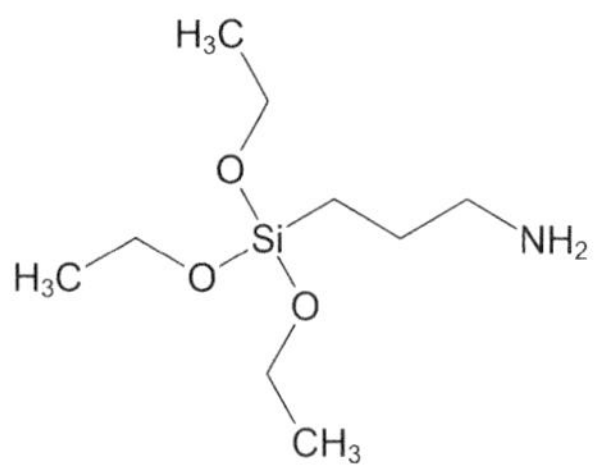

(3-Aminopropyl)triethoxysilane

Figure 3.13 Adhesion aids (20).

The silicone rubber sheet may be configured to have a cylindrical shape in such a way as to cover the spouting nozzle of the inflator and be fixed to the airbag or the inflator. Also, the silicone rubber sheet may be disposed between the airbag and the inflator in such a way as to constitute at least part of the flow passage (20).

According to the silicone rubber sheet used for the airbag device, gas can be sufficiently fed into the airbag by covering the outer circumference of the spouting nozzle of the inflator with the silicone rubber sheet having certain strength and thickness in combination when the inflator is inserted into the airbag.

Therefore, a flow passage component, which has previously been produced by complicated forming of a metal pipe in the related art, is not necessary anymore, so that miniaturization and weight reduction can be facilitated around the inflator. This will contribute to the cost reduction. Meanwhile, the silicone rubber sheet is stacked on the fiber fabric and, therefore, expansion and deformation of the silicone rubber sheet when the gas is fed from the inflator can be suppressed (20).

3.1.3 Silicone Rubber Sponge

Owing to its heat resistance, light weight, and superior weatherability, silicone rubber sponges are used for automotive components, copier roll surface-covering materials, and for various sealing materials (21).

A sponge-formable silicone rubber composition contains (22,23):

1. 100 parts by mass of a poly(organo siloxane) having at least two silicon-bonded alkenyl groups per molecule,
2. A poly(organo siloxane) having at least two silicon-bonded hydrogen atoms per molecule, in an amount such that an amount of silicon-bonded hydrogen atoms in component 2 is from 0.4 to 20 moles per 1 mole of alkenyl groups in component 1,
3. From 20 to 500 parts by mass of water,
4. Cellulose nanofibers, in an amount of from 0.01 to 15 parts by mass per 100 parts by mass of component 3,
5. From 0.1 to 15 parts by mass of an emulsifier, and
6. A hydrosilylation reaction catalyst, in an amount to accelerate a hydrosilylation reaction of the composition.

The composition may also contain an electrically conductive filler for the purpose of imparting electrical conductivity to the obtained silicone rubber sponge. Examples of an electrically conductive filler include carbon-based conductive agents such as carbon black, carbon fibers, carbon nanotubes, and graphite; also metal powders such as gold, silver, and nickel, and electrically conductive zinc oxide.

The silicone rubber sponge is obtained by crosslinking the composition by a hydrosilylation reaction and then removing the water from the obtained silicone rubber, or by removing the water during crosslinking by a hydrosilylation reaction.

An example for the preparation is as follows (22, 24):

Preparation 3–5: First, 100 parts by mass of poly(dimethyl siloxane) capped at both molecular terminals with dimethylvinylsiloxy groups having a viscosity of 40,000 *mPa s*, 50 parts by mass of fumed silica having a BET specific surface area of 400 $m^2 g^{-1}$, 10 parts by mass of hexamethyldisilazane, 0.36 parts by mass of tetramethyldivinylsilazane, 2 parts by mass of water, and 0.26 parts by mass of dimethyl siloxane-methylvinyl siloxane copolymer capped at both molecular terminals with dimethylhydroxysiloxy groups having a viscosity of 20 *mPa s* (vinyl group content approximately 10.9%) were loaded into a Ross mixer, and after mixing until uniform at room temperature, it was processed while heating at 200°C under reduced pressure for 2 *h* to prepare a fluid silica master batch.

The sponge-formable silicone rubber composition is able to form a silicone rubber sponge having uniform and fine bubbles which does not readily contaminate a mold (22).

Since the sponge-formable silicone rubber composition has an excellent moldability and can form a silicone rubber sponge having fine bubbles after molding, it is advantageously used in elastic materials of rollers and belts used in image-forming devices such as electronic photographic copiers, laser printers, on-demand printers, and facsimile machines; and sponge applications used at high temperature such as thermal insulation materials, sound absorbing materials, cushioning, packing, gaskets, and pads (22).

3.1.3.1 Conductive Silicone Rubber Sponge

Conventionally known conductive silicone rubber sponges are formed by blending carbon black into a silicone rubber as a conductive material to control the electrical resistance and further adding a foaming agent and curing by heating (25).

Since this conductive silicone rubber sponge has excellent heat resistance and weather resistance and is light weight, the sponge is used in a wide range of applications as an automotive part or an office equipment part. Such sponges are used as various sealing materials, packing, gaskets, O-rings, or roll coating materials.

A conventional method used for forming such a conductive silicone rubber sponge is a method of blending a pyrolytic organic foaming agent, such as 2,2′-azobisisobutyronitrile, but this method has problems in that the characteristics of the sponge molded product are diminished by the organic foaming agent decomposition residue, the decomposition gas itself is toxic or has an odor, and curing can be inhibited by the organic foaming agent when a platinum catalyst is used as a curing agent. Furthermore, the change in the resistance when the applied voltage is varied, i.e., the voltage dependence of the resistance is large, which makes it difficult to obtain a conductive member with stable electrical resistance characteristics.

A conductive silicone rubber sponge-forming composition has been developed that is capable of forming a conductive silicone rubber sponge with a fine, uniform, and continuous porosity, where the voltage dependence of the electrical resistance is low (25).

The method of producing the silicone rubber sponge is done using the following steps (25):

Step 1: This step prepares a sponge-forming liquid silicone rubber composition comprising at least the following components (A) to (G). The components (A) to (H) in Step 1 are described in Table 3.5. The hardness or physical properties of the resulting conductive silicone rubber sponge can be changed as necessary by changing the compounded amount of component (B). The volume resistivity of the resulting conductive silicone rubber sponge can be changed as necessary by changing the compounded amount of component (C). The density of the resulting conductive silicone rubber sponge can be changed as necessary by changing the compounded amount of component (D).

Step 2: This step obtains a silicone rubber molded body in a hydrated state by crosslinking and curing the conductive sponge-forming liquid silicone rubber composition obtained in Step 1.

Step 3: This step obtains a conductive silicone rubber sponge by removing the water from the silicone rubber molded body in a hydrated state obtained in Step 3. The method of removing water from the silicone rubber molded body in a hydrated state in Step 3 of the method of producing the silicone rubber sponge is not particularly limited, but the water can be preferably removed by heating the silicone rubber molded body in a hydrated state prepared in Step 2 at a temperature of from 120°C to 250°C.

In addition, the conductive sponge-forming liquid silicone rubber composition described above preferably further contains a fine silica

Table 3.5 Components of sponge-forming liquid silicone rubber (25).

No.	Description
(A)	100 Parts by mass of an alkenyl group-containing poly(-diorgano siloxane) composed of:
(A1)	From 0 to 90 parts by mass of a poly(diorgano siloxane) having both molecular terminals capped with alkenyl groups and not having any alkenyl groups in molecular side chains, and
(A2)	From 10 to 100 parts by mass of a poly(diorgano siloxane) having two or more alkenyl groups in molecular side chains
(B)	A poly(organo siloxane) having at least two silicon-bonded hydrogen atoms in a molecule, in an amount such that the number of moles of silicon-bonded hydrogen atoms in component (B) is from 0.4 to 20 per 1 mole of alkenyl groups in component (A)
(C)	From 1 to 100 parts by mass of a conductive filler
(D)	From 10 to 250 parts by mass of a mixture comprising water and an inorganic thickener
(E)	From 0.1 to 15 parts by mass of an emulsifier
(F)	A hydrosilylation reaction catalyst, in an amount sufficient for crosslinking the composition
(G)	From 0.001 to 5 parts by mass of a curing retarder

powder (H) within the range of from 0.01 to 20 parts by mass per 100 parts by mass of component (A) from the perspective of improving the strength of a conductive silicone rubber sponge obtained by the production method. Component (H) is as described above (25).

A fluorine resin layer or a fluorine rubber layer may be further used on the outer periphery of the conductive silicone rubber sponge layer. In this case, a fluorine resin layer is formed by a fluorine resin coating material or a fluorine resin tube to coat the conductive silicone rubber sponge layer. Examples of fluorine resin coating materials include latex of a PTFE, a perfluoroalkyl vinyl ether copolymer resin, a fluorinated ethylene-propylene copolymer resin (FEP), a poly(vinylidene fluoride) resin, and a polyfluorinated vinyl resin (25).

3.1.4 *Dilatant Fluid*

A dilatant fluid is a mixture of liquid and particles and behaves as a fluid under slow deformation and as a solid in response to rapid deformation (26). Recently, dilatant fluids were used for protecting a fuel cell stack mounted inside fuel cell vehicles. In the fuel cell stack, the dilatant fluid conforms to slow deformation, as in the case of cell deflection, to prevent the cells from being further deformed or damaged. On the other hand, when large deformation is applied, as in the case of a minor collision, the dilatant fluid behaves as a solid to prevent misalignment of the cell stack.

However, a problem may occur when the dilatant fluid is received in a bag and the dilatant fluid leaks out upon breakage of the bag. If a solid material like rubber exerts dilatant properties, such a problem would be overcome (26).

A silicone composition has been developed which has a low viscosity and good moldability prior to curing and cures into a dilatant product that exhibits a low storage elastic modulus at a low strain rate and a high storage elastic modulus at a high strain rate (26).

It has been found that when an organopolysiloxane component capped at one end with an alkoxysilyl group, a linear organohydrogenpoly(siloxane) component having Si–H groups on side chains, and a linear organohydrogenpolysiloxane component having Si–H groups at both ends are used in a filled addition-curable silicone composition, the resulting silicone composition has a low viscosity despite the filler loading, is easy to handle prior to curing, and cures into a cured product that exhibits a low storage elastic modulus when the strain rate is in a low-frequency range, i.e., 0.1 *Hz*, and a high storage elastic modulus when the strain rate is in a high-frequency range of 500 *Hz*.

The components used for such compositions are shown in Table 3.6. The preparation runs as follows (26):

Table 3.6 Components used (26).

Number	Content
A-1	Dimethylpolysiloxane capped at both ends with dimethylvinylsilyl, having a viscosity of 5.0 *Pa s* at 25°C
A-2	Dimethylpolysiloxane capped at both ends with dimethylvinylsilyl, having a viscosity of 10.0 *Pa s* at 25°C
A-3	Dimethylpolysiloxane capped at both ends with dimethylvinylsilyl, having a viscosity of 0.4 *Pa s* at 25°C
B	Organopolysiloxane
C	Organohydrogenpolysiloxane
D	Organohydrogenpolysiloxane
E	Filler: Spherical alumina powder having an average particle size of 4.0 μm and a specific surface area of 0.50 $m^2 g^{-1}$
F	Platinum group metal catalyst, platinum-divinyltetramethyldisiloxane complex
G	Reaction inhibitor, 1-Ethynyl-1-cyclohexanol

Preparation 3–6: A 5-l planetary mixer (Inoue Mfg. K.K.) was charged with components (A), (B) and (E), which were mixed at 25°C for 2 *h* under reduced pressure. Component (F) was added thereto and mixed at 25°C for 30 *min*. Then, component (G) was added thereto and mixed at 25°C for 30 *min* until uniform. Finally, components (C) and (D) were added thereto and mixed at 25°C for 30 *min* until uniform, yielding the final silicone composition.

3.1.5 *Thermally Conductive Adhesive Composition*

Since heat-generating parts, such as power transistors and thyristors, degrade their characteristics by the heat generated, a countermeasure is taken upon installation by attaching a heat sink to the heat-generating part so that the heat may be dissipated and released to a metallic chassis of the apparatus (27). In this case, a heat-dissipating insulating sheet of silicone rubber loaded with heat-conductive filler is used between the heat-generating part and the heat sink for the purpose of improving electrical insulation and thermal conductivity.

When a composition is heavily loaded with a heat-conductive filler for increasing its thermal conductivity, the fluidity is reduced and molding or working ability is extremely exacerbated (27).

A heat-conductive silicone adhesive composition has been described (27). It can be used with reactors, even when heavily loaded with a heat-conductive filler. The composition is highly heat-conductive, maintains good flow, can be potted around delicate substrates, has good physical properties after curing, experiences little changes of physical properties even with heat or moist-heat aging, and has good adhesion to metals and organic resins, and a reactor potted in the composition.

The heat-conductive silicone adhesive composition contains an organopolysiloxane, c.f. Figure 3.14.

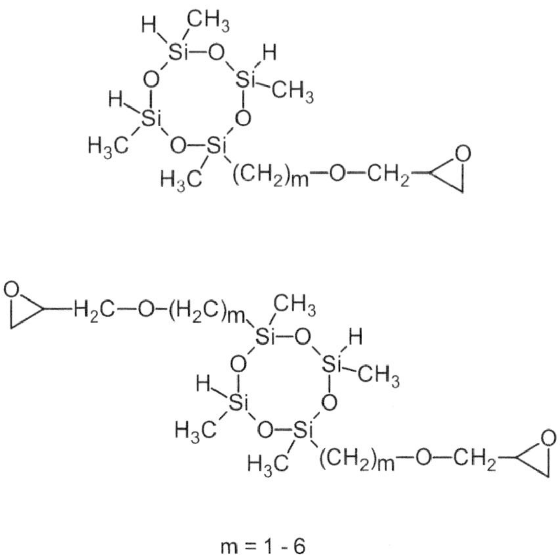

Figure 3.14 Organopolysiloxane.

Furthermore, a heat-conductive filler is added to increase the thermal conductivity of the composition. The heat-conductive filler is preferably an inorganic powder and/or a metal powder. A hydrosilylation reaction catalyst is added in order to promote the curing of the composition, preferably a platinum group metal-based catalyst.

An organopolysiloxane is used that contains at least one silyl group of the general formula $SiR^1_aR^2_{3-a}$. Here R^1 is a substituted or unsubstituted monovalent hydrocarbon group, R^2 is an alkoxy or acyloxy group of 1 to 8 carbon atoms, and a is 0, 1 or 2. The blending of this component ensures that the composition is easy to handle and mold and more self-adhesive to metals, glass and organic resins even when heavily loaded with the heat-conductive filler. To enhance the thermal conductivity of the composition, acetylene black, obtained from pyrolysis of acetylene, is added.

Also, a reactor potted in the composition is provided. This ensures the reliable performance of the reactor for applying the battery voltage across the motor after boosting up.

As shown in Figure 3.15, the reactor is constructed by winding an insulated copper wire around a dust core of magnetic material to form a coil body, placing the coil body in a metal case of aluminum or the like, potting or cast molding the composition into the space so that the coil body is embedded in the composition, and curing the composition.

The components, including coil body and metal case, the dispenser for potting, and the potting method used herein, may be selected from well-known components and technologies.

The reactor was heat aged by heating at 170°C for 1,000 *h*, and also moist-heat aged by heating at 85°C and 85% relative humidity for 1,000 *h*. The electricity is conducted across each aged reactor in the same conduction pattern as at the initial pattern, while the temperatures (thermal properties after heat aging and moist-heat aging) of coil body and potted material were measured.

Then a temperature difference (change of reactor performance) between the initial thermal properties and the heat or heat-moist aged thermal properties was computed (27).

It could be demonstrated that silicone adhesive compositions containing a liquid organohydrogenpolysiloxane with at least 2 silicon-bonded hydrogen atoms in the molecule and containing an alkoxy group, the polysiloxane having a cyclic structure-containing

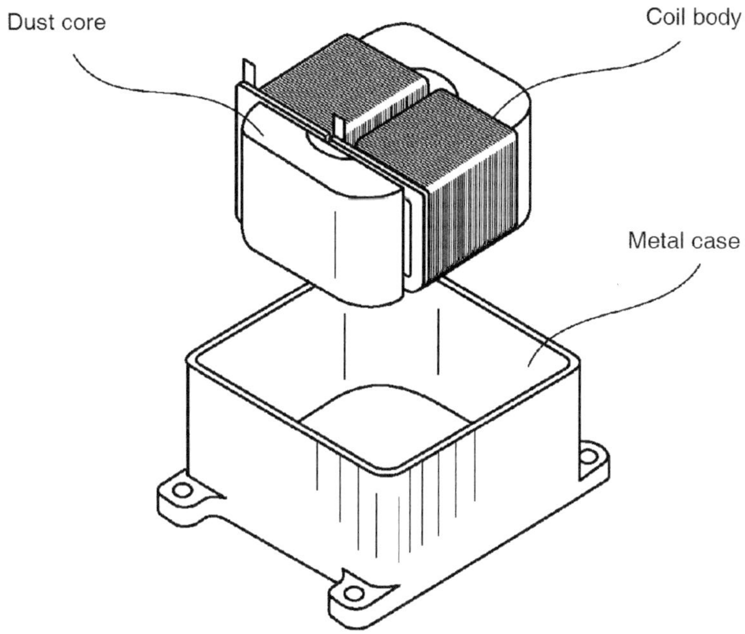

Figure 3.15 Reactor (27).

skeleton, undergo substantial changes of physical properties after moist-heat treatment relative to the initial physical properties (27).

3.1.6 Automobile Exhaust Systems

The use of silicone rubber as anti-vibration rubber in automobile exhaust systems is known to be an effective general countermeasure to heat damage (28). However, there are limits for using a silicone rubber because its vibration transmission characteristics are not as good as those of ethylene propylene diene monomer (EPDM), a rubber material commonly used in exhaust mounts.

A new silicone rubber material with better vibration transmission characteristics and lower damping than EPDM was developed using a liquid silicone rubber (28). A technique that reduces the damping components of rubber was used to construct low damping to anti-vibration rubber materials like EPDM. Such damping components include reinforcing agents such as carbon black and silica, which are injected to reinforce the rubber.

Damping in the polymer is the result of internal friction when rubber deforms, so damping is controlled by using a material with only a few side chains. Damping in the filler is the result of deformation of the aggregates that form, so damping is controlled by using a grade of materials with large particles and small, secondary aggregate structures.

Until now, there have been limits on how much one could lower damping by using these conventional techniques with silicone rubber. Among other reasons, there are few variations on dry silica, which needs to be used in the reinforcing agent. Thus, this research used liquid silicone rubber and successfully controlled the friction between polymers, providing low damping by achieving a high-density crosslinking structure. It could be confirmed that components made with such materials offered better vibration transmission characteristics in comparison to EPDM components with the same shape and static spring value.

The spring properties of silicone rubbers show only a little dependence on temperature, thus simultaneously enhancing the vibration transmission characteristics in the low temperature range (28).

3.2 Underwater Vehicles

3.2.1 Buoyancy Control Device

A small buoyancy control device with liquid silicone rubber for underwater vehicles has been described that utilizes the volume difference of material between solid and liquid states to induce a buoyancy change (29).

The stretching characteristics of several materials were compared in terms of specific gravity and heat-proof temperature. Three types of buoyancy control devices were manufactured that contained a solid or liquid paraffin wax enclosed by silicone rubber. A sketch of such a buoyancy control device is shown in Figure 3.16.

Also, two types of heating equipment for the developed buoyancy control device were built through which hot water flowed and it was found that the buoyancy change of one of the types of equipment used roughly agreed with the theoretical value (29).

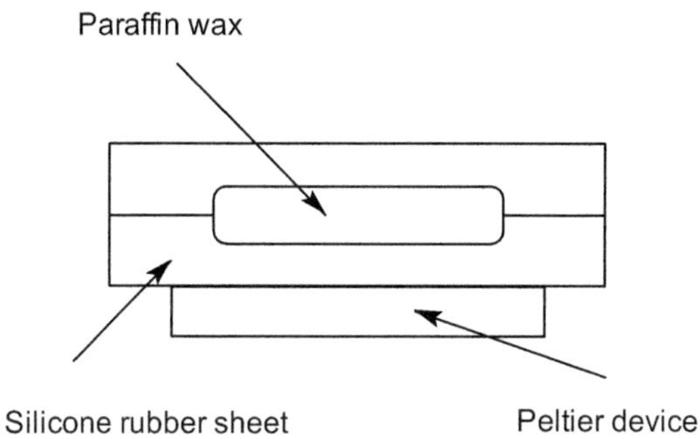

Figure 3.16 Buoyancy control device (29).

References

1. S. Wang, Method for the adhesion of a layer of fluorosilicone rubber to a layer of silicone rubber, US Patent 9 908 308, assigned to Dow Corning (China) Holding Co., Ltd. (Shanghai, CN), March 6, 2018.
2. J.W. Hetrick, Safety cushion assembly for automotive vehicles, US Patent 2 649 311, assigned to John W. Hetrick, February 5, 1953.
3. W. Linderer, Means for the protection of in-vehicle persons against injuries in clashes, DE Patent 896 312, assigned to Walter Linderer, June 12, 1953.
4. Wikipedia contributors, Airbag — Wikipedia, the free encyclopedia, https://en.wikipedia.org/w/index.php?title=Airbag&oldid=874243791, 2018. [Online; accessed 15-January-2019].
5. G.J. Humphreys, A.K. Breed, and T. Thuen, Air bag assembly for motor vehicles, US Patent 5 348 340, assigned to Breed Automotive Technology, Inc. (Lakeland, FL), September 20, 1994.
6. K. Nakamura, S. Miyata, M. Hotta, T. Yoneyama, and M. Fukuda, Silicone rubber sheet and airbag device, US Patent 9 352 719, assigned to Takata Corporation (Tokyo, JP), Shin-Etsu Chemical Co., Ltd. (Tokyo, JP), May 31, 2016.
7. M. Barth, W. Blackwood, V. Clerici, and A.W. Mountney, Coated fabric products, US Patent Application 20 180 313 030, assigned to Dow Silicones Corp., November 1, 2018.
8. S. Yamamoto, Y. Tsuji, and W.R. Blackwood, Liquid curable silicone rubber composition and woven fabric coated with cured product of the same composition, US Patent 8 962 498, assigned to Dow Corning Toray Co., Ltd. (Chiyoda-ku, Tokyo, JP), Dow Corning Corporation (Midland, MI), February 24, 2015.
9. T. Nozoe, Y. Tsuji, and S. Yamamoto, Silicone rubber composition for coating for woven fabric, and coated woven fabric, US Patent 9 062 411, assigned to Dow Corning Toray Company, Ltd. (Chiyoda-ku, Tokyo, JP), June 23, 2015.
10. H. Mizushima, T. Kimura, and S. Ubukata, Liquid silicone rubber coating composition, curtain airbag, and its production method, US Patent 8 785 586, assigned to Shin-Etsu Chemical Co., Ltd. (Tokyo, JP), July 22, 2014.
11. FMVSS, Flammability of interior materials – passenger cars, multipurpose passenger vehicles, trucks, and buses, FMVSS 302, Federal Motor Vehicle Safety Standard, USA, 1972.
12. ISO, Road vehicles, and tractors and machinery for agriculture and forestry – determination of burning behaviour of interior materials, ISO 3795, Federal Motor Vehicle Safety Standard, Geneva, 1989.

13. O. Hayashida, T. Uno, K. Oishi, and A. Kashima, Fluorosilicone rubber composition, US Patent 9 777 157, assigned to Shin-Etsu Chemical Co., Ltd. (Tokyo, JP), October 3, 2017.
14. JIS, Standard test methods for liquid-contaminant, inclined-plane tracking and erosion of insulating materials, JIS K 6249:2003, Japanese Standards Association, Tokyo, JP, 2003.
15. O. Hayashida and A. Kashima, Fluorosilicone rubber compound and rubber part for use near cargo plane engines, US Patent 9 458 306, assigned to Shin-Etsu Chemical Co., Ltd. (Tokyo, JP), October 4, 2016.
16. Japanese Standards Association, Rubber, vulcanized or thermoplastic – determination of compression set at ambient, elevated or low temperatures, Online, electronic: https://global.ihs.com/doc_detail.cfm?document_name=JIS%20K%206262&item_s_key=00256585 JIS K 6262, Japanese Standards Association, Tokyo, JP, 2013.
17. H. Fujiki, Y. Inoue, Y. Harada, T. Miyao, and M. Ikeno, Liquid silicone rubber coating composition and air bag fabric, US Patent 6 387 520, assigned to Shin-Etsu Chemical Co., Ltd. (Tokyo, JP), May 14, 2002.
18. Y. Tsuji and A. Nakamura, Silicone rubber based coating composition for air bags, US Patent 6 420 037, assigned to Dow Corning Toray Silicone Co. Ltd. (Tokyo, JP), July 16, 2002.
19. A. Nakamura and Y. Tsuji, Liquid silicone rubber coating composition, US Patent 5 789 084, assigned to Dow Corning Toray Silicone Co., Ltd. (Tokyo, JP), August 4, 1998.
20. K. Nakamura, S. Miyata, M. Hotta, T. Yoneyama, and M. Fukuda, Silicone rubber sheet and airbag device, US Patent Application 20 150 123 387, assigned to Shin-Etsu Chemical Co., Ltd, Tokyo, (JP) and Takata Corporation, Tokyo, (JP), May 7, 2015.
21. S. Shirasaki, T. Fukamachi, and M. Hamada, Emulsion composition for silicone rubber sponge, process for producing the same, and process for producing silicone rubber sponge, US Patent 8 227 520, assigned to Dow Corning Toray Company, Ltd. (Chiyoda-ku, Tokyo, JP), July 24, 2012.
22. M. Irie, Sponge-formable silicone rubber composition and silicone rubber sponge, US Patent Application 20 180 037 709, assigned to Dow Corning Toray Co. Ltd., February 8, 2018.
23. M. Irie, Sponge-formable silicone rubber composition and silicone rubber sponge, US Patent Application 20 180 057 652, assigned to Dow Corning Toray Co. Ltd., March 1, 2018.
24. T. Nozoe, F. Suto, H. Yoshida, A. Sakuma, and Y. Tsuji, Multi-component sponge-forming liquid silicone rubber composition and silicone rubber sponge manufacturing method, US Patent 9 051 445, assigned to Dow Corning Toray Co., Ltd. (Chiyoda-ku, Tokyo, JP), June 9, 2015.
25. H. Akitomo, M. Irie, and H. Kurusu, Conductive sponge-forming silicone rubber composition and conductive silicone rubber sponge,

US Patent Application 20 150 228 372, assigned to Dow Corning Toray Co., Ltd., Tokyo (JP), August 13, 2015.
26. N. Matsumoto, A. Sakamoto, N. Goto, M. Ishii, H. Takeuchi, Y. Suzuki, and S. Miyoshi, Silicone composition and cured product, US Patent Application 20 180 179 384, assigned to Shin-Etsu Chemical Co., Ltd., Tokyo (JP), Toyota Jidosha Kabushiki Kaisha, Toyota-shi (JP), and Soken, Inc., Nishio City (JP), June 28, 2018.
27. M. Iwata, M. Kobayashi, K. Nakanishi, H. Imanishi, and H. Yamazaki, Thermally conductive silicone adhesive composition for reactor and reactor, US Patent 9 424 977, assigned to Shin-Etsu Chemical Co., Ltd. (Tokyo, JP), Toyota Jidosha Kabushiki Kaisha (Toyota-shi, JP), August 23, 2016.
28. A. Kaneta, Development of exhaust mount which applied low damping liquid silicone rubber, in *SAE Technical Paper*, number 156, pp. 1–5. SAE International, April 2018.
29. H. Yamamoto and K. Shibuya, New small buoyancy control device with silicone rubber for underwater vehicles, in M.O. Tokhi and G.S. Virk, eds., *Advances in Cooperative Robotics*, pp. 258–265, Singapore, 2017. 19th International Conference on CLAWAR 2016, World Scientific.

4

Electrical and Optical Uses

4.1 Electrically Conductive Silicone Rubber

The application of elastomeric insulating materials in high voltage components has significantly increased due to their valuable electrical and mechanical properties (1). However, although knowledge concerning the DC conductivity of these materials is limited, it is required prior to implementation of high voltage current insulation.

With the objective of fostering knowledge on DC conductivity of liquid silicone rubber (LSR), measurements were performed using plaque samples. The current-time behavior was measured when a field strength of 10 $kV\,mm^{-1}$ and a temperature of 45°C was applied (1).

The so-obtained results provided the base for statistically covered conclusions on the determination of DC conductivity of LSR. Influences of voltage polarities are analyzed and dependencies on the measured conductivity, related to previously applied voltage stresses, as known from XLPE, i.e., crosslinked poly(ethylene), were discussed.

It was concluded that for the investigated LSR no significant influences on the measured conductivity due to short circuit duration, polarity, orientation or previously applied voltage stresses can be found (1).

4.1.1 *Conductive Liquid Silicone Rubber-Based Composites*

4.1.1.1 *Multiwalled Carbon Nanotubes*

Conductive liquid silicone rubber-based composites were prepared through the incorporation of fumed silica and poly(dimethyl siloxane) modified multiwalled carbon nanotubes (CNTs) (2). Their mechanical and electrical properties were investigated. It was shown that the introduction of poly(dimethyl siloxane) modified multiwalled CNTs and fumed silica with a ratio of 0.25/30 increases the crosslinking density and the thermal stability of the composites significantly.

The improved mechanical performance, e.g., 60% increase of Young's modulus and a 47% increase for tear strength, should be ascribed to the stronger interfacial interaction between the CNTs and poly(dimethyl siloxane) matrix and better dispersion of the composites due to the existence of poly(dimethyl siloxane) on the CNT surface. Moreover, the better dispersion of the composites also ensures a much higher electrical conductivity than original carboxyl functionalized CNT at a relatively low loading, which could be attributed to the formation of a CNT network (2).

4.1.1.2 *Fibrous Carbon Nanostructures*

Silicone rubber compositions containing a silicone rubber blended with carbon nanotubes and vulcanized products of the compositions have been required to show high electrical conductivity while remaining flexible like rubber (3).

However, when a silicone rubber is blended with large amounts of fibrous carbon nanostructures, such as carbon nanotubes, for increased electrical conductivity, the silicone rubber composition and vulcanized product show increased hardness and thus low flexibility, resulting in poor adhesion and shape conformability during use.

Silicone rubber compositions have been developed that contain a silicone rubber and fibrous carbon nanostructures including carbon nanotubes, wherein the fibrous carbon nanostructures exhibit a convex upward shape in a t-plot obtained from an adsorption isotherm. When such fibrous carbon nanostructures that exhibit a convex upward shape in a t-plot obtained from an adsorption isotherm are blended, it is possible to sufficiently increase the electrical conductivity of the silicone rubber compositions while limiting rises in hardness (3).

The fibrous carbon nanostructures preferably have an average diameter of 2 *nm* to 10 *nm*. When fibrous carbon nanostructures having an average diameter of 2 *nm* to 10 *nm* are used, higher levels of flexibility and electrical conductivity can be attained at the same time. The vulcanized product should preferably have a hardness of 80 or less. When the hardness is 80 or less, sufficient adhesion and shape conformability can be ensured during its use (3).

4.1.2 *Effect of Shape and Size of Nickel-Coated Particles on Conductivity*

The electrical resistivity of the composites is influenced by several factors (4).

In particular, the resistivity of the conductive rubber is mainly influenced by the concentration, size, shape, and structure of filler particles, the type of matrix, and the processing conditions (5).

The conductive silicone rubbers (CSRs) were prepared by adding the nickel-coated graphite and/or nickel-coated carbon fiber into liquid silicon rubbers, which were then subjected to vulcanization.

Poly(methyl hydrogen siloxane) was used as crosslinking agent and 1-ethynyl-1-cyclohexanol was used as retarder. The nickel-coated graphite contained 75% of Ni, the width-to-thickness ratio was 5:1 to 15:1 and average thickness was 5 μm. The nickel-coated carbon fiber contained 66% wt of Ni, with a length-to-diameter ratio of 4:1–10:1 and the average diameter was 7 μm.

The effect of particle shape and size on the electrical conductivity of CSR was investigated (4). The results indicate that nickel-coated carbon fiber-filled CSR have lower percolation threshold than nickel-coated graphite-filled CSR. Compared to the filled particles with larger size, a higher amount of smaller particles are needed to form the same conductive pathway in CSR.

Thus, there are more contact points in the conductive pathway, which increase the total contact resistance. The volume resistivity of CSR can be significantly reduced by the doping of nickel-coated carbon fiber particles, while the effect begins to wear off under high doping content.

The fiber-like nickel-coated carbon fiber has a good bridging function for the formation of conductive pathway in CSR, which is greatly enhanced by increasing the length-to-diameter (l/d) ratio of nickel-coated carbon fiber particles. Therefore, doping of the particles that have low percolation threshold can reduce the resistivity of the composites (4).

4.2 High-Voltage Insulation

An urgent topic in the field of high-voltage insulation is how to effectively improve the tracking and erosion resistance of a silicon rubber (SR) (6,7).

4.2.1 *Platinum Catalyst and Nitrogen-Containing Silane*

The tracking and erosion resistance of SR could be significantly improved by incorporating a platinum catalyst and a nitrogen-containing silane. The suppression effect and mechanism of Pt/nitrogen-containing silane on tracking and erosion were studied by inclined plane test, thermogravimetric analysis (TGA), thermogravimetric Fourier transform infrared analysis (TG-FTIR), laser Raman spectroscopy, and scanning electron microscopy (SEM) (6). These measurements revealed that when 1.4 *phr* of nitrogen-containing silane and 6.7 *ppm* of Pt were added, the tracking resistance of SR was improved from a 2.5 *kV* to 4.5 *kV* level in the inclined plane test, and the eroded mass was significantly reduced. This might be attributed to the synergistic effect of Pt/nitrogen-containing silane on silicone chains.

At a high temperature produced by arc discharge, Pt/nitrogen-containing silane would catalyze a radical crosslinking, meanwhile suppressing oxidation and depolymerization of silicone chains. Hence, a tightly crosslinked network was formed and protected inner materials from arc ablation. Moreover, the deposit of carbon during pyrolysis was suppressed by the Pt/nitrogen-containing silane, which served as the secondary mechanism of tracking suppression (6).

4.2.2 *Amine-Containing MQ Silicone Resin*

The MQ silicone resins consist of hydrolytic condensation products of monofunctional silane (M) and tetrafunctional silane (Q). An amine-containing MQ silicone resin (A-MQ) was prepared to enhance the tracking and erosion resistance of addition-cure liquid silicone rubber (ALSR) (7).

The preparation of an amine-containing MQ silicone resin was done as follows (7):

Preparation 4–1: In a 250 *ml* four-neck flask, 17.8 *g* of MM and 2.2 *g* of MViMVi were added to a solution composed of 10.8 *g* of HCl, 10.0 *g* of ethanol, and 14.4 *g* of deionized water, and the reaction was heated at 70°C for 30 *min* with stirring. Then, 41.6 *g* of tetraethoxysilane was added dropwise into the flask for 3 *h*, followed by stirring for an additional 30 *min*. Then, 1.2 *g* of *N*-(β-aminoethyl)-γ-aminopropyltriethoxysilane, c.f. Figure 4.1, was added dropwise to the solution under stirring for 1 *h* at 70°C. After the reaction was finished, 80 *ml* of toluene was added to the solution and mixed well.

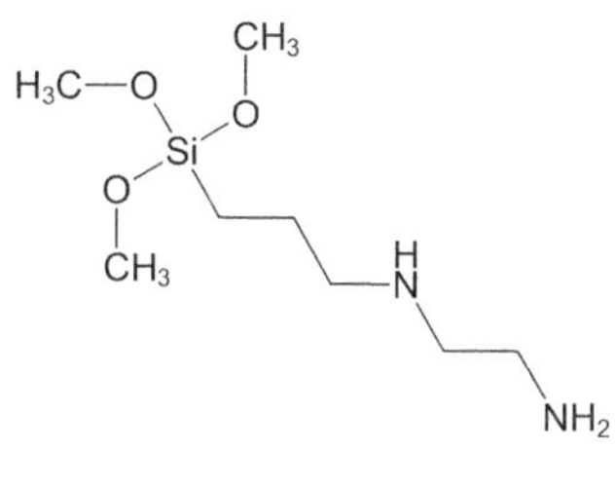

Figure 4.1 *N*-(β-Aminoethyl)-γ-aminopropyltriethoxysilane.

Then, the organic layer was separated, neutralized with $NaHCO_3$, dried with $MgSO_4$, and filtered. By removing the solvent under vacuum, A-MQ was obtained as a faint yellow viscous liquid. The vinyl group content and nitrogen content in A-MQ was 2.17% and 0.74%, respectively, which was determined by iodometric titration and hydrochloric titration.

The results of characterization showed that A-MQ imparted ALSR with excellent tracking and erosion resistance. When A-MQ content was 4 *phr*, all test samples passed the inclined plane test at 4.5 *kV*, and the erosion mass decreased by 67.8%. In addition, the tensile strength and tear strength increased by 13.2% and 13.6%, respectively, in comparison to that of an ALSR without A-MQ.

The effect of A-MQ content on the mechanical properties of the ALSR samples is shown in Table 4.1.

Table 4.1 Effect of A-MQ content on the mechanical properties of the ALSR samples (7).

Content /[phr]	Tensile strength /[MPa]	Elongation at break/[%]	Tear strength /[$kN\,m^{-1}$]	Hardness /[Shore A]
0	7.6 ± 0.3	638 ± 56	31 .6 ± 4.1	41 ± 1
1	8.1 ± 0.2	712 ± 38	33 .4 ± 3.6	42 ± 1
2	8.5 ± 0.4	698 ± 51	35 .1 ± 3.2	43 ± 1
3	8.9 ± 0.2	703 ± 60	36 .7 ± 5.3	44 ± 1
4	8.6 ± 0.3	678 ± 75	35 .9 ± 3.0	44 ± 1

Aspects of the suppression mechanism in heat attack and plasma bombardment were investigated by laser Raman spectroscopy, TGA, thermogravimetric FTIR analysis, laser Raman spectroscopy, and SEM, attenuated total reflection Fourier transform infrared (FTIR), and X-ray photoelectron spectroscopy.

This revealed that at the elevated temperature caused by arc discharge, A-MQ promoted the crosslinking of the polysiloxane molecules and suppressed the generation of cyclic oligomers, which reduced the intensity of the electrical arc. Moreover, when suffering from plasma bombardment, which was also produced by arc discharge, A-MQ protected the silicone chains from degradation and eliminated the carbon deposited on the surface (7).

4.2.3 *Tracking and Erosion Requirements*

One of the important disadvantages of silicone rubber insulators is the aging caused by their exposure to pollution and environmental factors such as UV, rain, and temperature (8).

The degradation can be highly dependent on the quality of the compounds used in the manufacturing of the insulators. In addition, the quality of the silicone compound depends strongly on the cost, the brand, production country, and others. On the other hand, for tracking evaluation, the IEC 62039 standard (9) recommends that for high voltage insulators, there is a minimum requirement of 3.5 *kV* and a satisfactory 2.5 *kV* for medium voltage applications. However, the IEC standard does not mention if this classification is valid for all environments, including the highly polluted ones, where tracking failures can develop (8).

Several failures occurred in two different 400 kV transmission lines, where polymeric insulators failed due to pollution and because of the quality and design of the materials. However, in the inclined plane tests done on the semicircular sheds of the insulators, most of the specimens did not surpass 4 kV to 4.75 kV. Thus, a tracking voltage should be established for silicone compounds exposed to polluted environments (8).

4.3 Silicone Rubber Composite Insulators

Silicone rubber composite insulators are now widely used in power systems (10–12).

Compared with glass and porcelain insulators, silicone rubber can deter leakage current and flashover due to the hydrophobicity of the material (13).

However, it also creates some new problems because of the organic nature of silicone rubber. These include surface damage induced by arc discharge (14–17).

The influence of superhydrophobicity on the surface damage of silicone rubber was studied (18).

A neat liquid one-component room temperature vulcanized silicone rubber was coated on a plate of glass. The dimensions of the glass were $25 \times 75\ mm^2$, with a thickness of 1 mm. The thickness of the coating was also about 1 mm. The samples were kept at room temperature for 24 h to allow curing. Common silicone rubber samples were acquired. Then, a mixture of ethyl acetate, liquid silicone rubber, and SiO_2 nanoparticles was sprayed onto the common silicone rubber samples to create superhydrophobic silicone rubber samples.

A droplet can become elongated on a common silicone rubber surface and arc discharge induced by the droplet can cause tracking on the silicone rubber surface. However, for a superhydrophobic silicone rubber surface, a droplet can leave the silicone rubber due to the low adhesion of the superhydrophobic surface.

Accordingly, arc discharge caused by the droplet does not occur, and the surface of the silicone rubber is not affected. Results demonstrate that using a superhydrophobic surface has a significant effect on limiting the surface damage of silicone rubber (18).

4.3.1 *Electrical Insulator*

A fast way of producing electrical insulations is to use room temperature vulcanizing silicone rubbers, including room temperature vulcanizing liquid silicone rubbers, which can cure in the presence of ultraviolet radiation.

The UV-curing of such UV vulcanizable silicone rubber compositions does not require or employ a heating step, and can occur within seconds to minutes. The use of UV vulcanizable RTV silicone rubber compositions, including room temperature vulcanizing liquid silicone rubbers, allows for the significant reduction of production times and therewith also production costs.

A method of producing a layer of a UV-cured silicone rubber composition on a substrate surface has been presented. The method consists of (19):

1. Applying a silicone primer composition to the substrate surface,
2. Hardening the silicone primer composition,
3. Applying a UV-curable silicone rubber composition on the silicone primer, and
4. UV curing the UV-curable silicone rubber composition to obtain an electrical insulator.

The silicone primer composition is suitable for promoting the adhesion of a vulcanized silicone rubber composition to a substrate surface.

Exemplary silicone primer compositions can be made from hydrolyzable silicates, hydrolyzable titanates, reactive hydrolyzable silanes and/or siloxanes.

Examples of compounds to be used as UV-sensitive crosslinking catalysts on the basis of palladium (Pd) are Pd(0)-compounds such as tetrakis(triphenylphosphino)palladium and the corresponding complexes with the ligands tri-(2-tolyl)phosphine, tris-(2-furyl)phosphine, tris(*tert*-butyl)phosphine, or the two-valent ligands 1,1-bis(diphenylphosphinomethane) or 1,2-bis-(diphenylphosphino)ethane. Some of these ligands are shown in Figure 4.2.

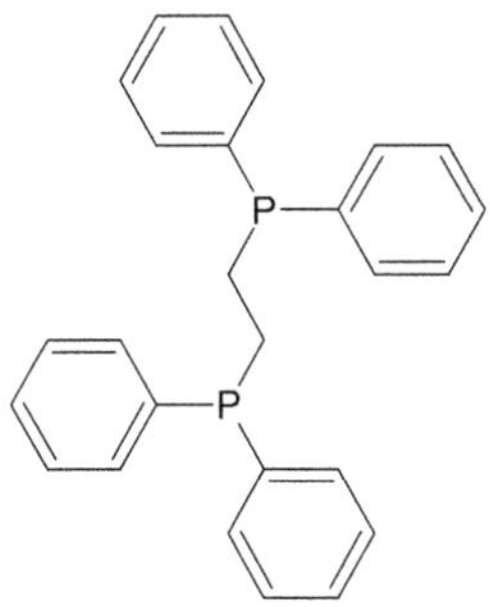

1,2-Bis-(diphenylphosphino)ethane

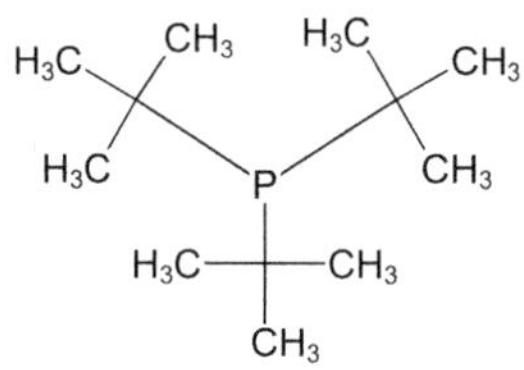

Tris(*tert*-butyl)phosphine

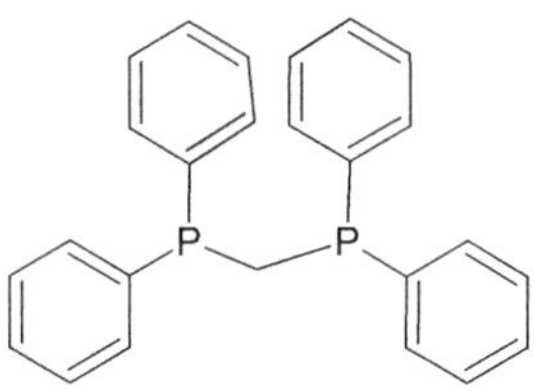

1,1-bis(Diphenylphosphinomethane)

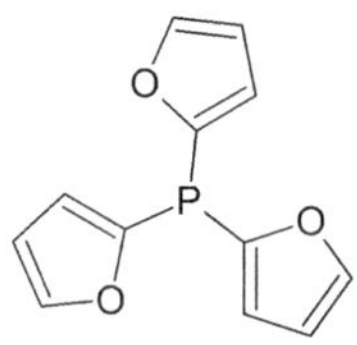

Tris-(2-furyl)phosphine

Figure 4.2 Ligands for catalysts (19).

4.3.1.1 Electrical Insulator for High Voltage Direct Current Applications

The transportation of electrical energy over the distances above 900 *km* is more efficient if high voltage direct current is used instead of alternating current (AC) (20). Therefore, the requirements for new cables and cable accessories in high voltage direct current reach up to 1000 *kV*. Insulating materials which are introduced in alternating current technology cannot be simply applied for high voltage direct current technology. The applicability of classical insulating materials in the increasing field of high voltage DC applications, in particular with increased voltage, is limited. For cable accessories like cable joints, materials like ethylene propylene diene monomer (EPDM) rubber and silicones are widespread, since they provide a high level of volume resistivity and desirable mechanical properties for the installation in place.

The electrical stress under AC and DC conditions differs significantly. Therefore, insulator materials useful for alternating current cannot be easily used for direct current as well as high voltage applications in particular (20).

Insulating materials with lower dielectric constant are exposed to higher stress under AC conditions, whereas materials with lower electrical conductivity are exposed to higher stress under DC conditions.

In order to overcome some problems of electrical stress and polarization inside of high voltage DC insulators, many proposals have been made. One group comprises multilayered materials, wherein each different layer has different dielectric constants or different resistivities. The majority of inventors and authors previously applied nonlinear or field gradient materials, e.g., in patent (21), where a volume resistivity in a range between 0.5 to 10 $kV\ mm^{-1}$ which decreases from 1012 to 108 Ωcm was disclosed.

Silicone compositions have been described that have a high electrical resistivity at a low temperature coefficient under high voltage of direct current conditions (20). It is a characteristic of these materials that electrical resistivity is on a higher level than for known field gradient materials although including the same base polymer. The preparation of a master batch was produced as follows (20):

Preparation 4–2: First, 100 *kg* of a vinyl-terminated linear poly(dimethyl siloxane) having a viscosity of 10 *Pa s* at 20°C was placed in a planetary mixer and mixed with 12.7 *kg* carbon black Ketjenblack EC-300J (Akzo) having BET surface 800 $m^2\,g^{-1}$ with a 3.50 DBP pore volume of $ml\,g^{-1}$ with a primary particle size of 40 *nm*. This mixture was stirred in a twin blade kneader till a homogeneous mixture was obtained after 45 *min*. The homogeneous mixture was then further dispersed over 30 *min* on a three-roll mill to obtain a much better dispersion of the carbon black. After this treatment, all particles in the filler batch show a particle size of smaller than 10 μm by a grindometer evaluation and the mixture had a viscosity at 20°C of 550-700 *Pa s* at a shear rate D=101 s^{-1}.

4.3.1.2 *Leakage Current*

Insulators generally are installed outdoors, so they can be affected by the weather conditions and the surrounding environment, e.g, the influence of moisture, rainwater, lightning strikes, temperature, or ultraviolet radiation (22). Some environmental issues may lead to a decrease in the working performance of the insulators to below its rating design. This may result in an insulation failure.

The so generated leakage current will also bring about power losses in the transmission line. Glass and ceramics are widely used as materials for insulators. This type of insulator has some disadvantages because of the changes in surface resistance under wet conditions. This may lead to a higher possibility of current leakage.

Actually, an attempt has been made to use a silicone-rubber polymer material for insulators, as it provides some advantages in comparison to ceramic and glass materials.

The effect of voltage and wetting conditions on the leakage current of a post-insulator made from a silicone rubber has been studied (22).

The preparation of silicone rubber materials can be done by vulcanization of a poly(dimethyl siloxane) polymer, covering the curing process to make the long molecules interlink with each other through the condensation process. The fabrication can be carried out at room temperature or high-temperature conditions using different catalysts.

In a clean state, the surface of an insulator has a large electrical resistance. The formation of a conductive layer on its surface is commonly caused by the contamination of water on the insulator surface. The conductive layer may cause the deterioration of the surface resistance, and furthermore may produce the leakage current occurrence on the surface.

The experimental setup is shown in Figure 4.3.

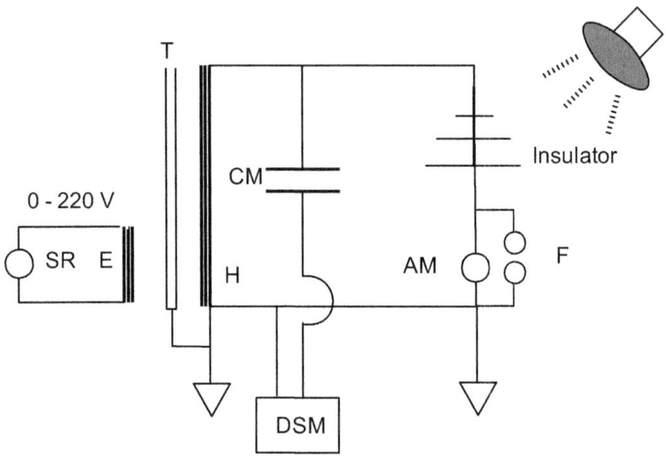

Figure 4.3 Experimental setup for leakage current observation (22).

Here, the symbol SR represents the slide regulator to adjust to input voltage, T is the testing transformer, E and H are the low-voltage and high-voltage windings of the transformer respectively, CM is the capacitor measurement, DSM is the instrument for measuring the peak AC high-voltage, AM represents the μ-amperemeter and F is the spark gap for overvoltage protection.

An alternating current (AC) has been used in the experiment to represent the actual conditions. The AC high-voltage applied to the silicon rubber insulator was generated from the testing transformer. It can be used to generate the voltage up to 100 kV_{rms}. The high-voltage level can be changed by adjusting the input voltage at the low-voltage winding E. The output voltage of 100 kV could be generated on the H winding when the low-voltage winding E is supplied with a voltage of 220 V. The range of voltage of 10 kV to 30 kV has been used in the course of the experiment.

The influence of the applied voltage on the occurrence of leakage current under dry conditions is shown in Table 4.2 and the influence of the applied voltage on the occurrence of leakage current under wet conditions is shown in Table 4.3. Also, a plot is shown in Figure 4.4.

Under dry conditions, the leakage current increases proportion-

Table 4.2 Applied voltage and leakage current under dry conditions (22).

voltage/[kV]	Average current/[μA]
10.06	9.27
15.06	14.27
20.05	19.20
25.06	24.27
30.05	29.37

Table 4.3 Applied voltage and leakage current under wet conditions (22).

Wetting/[$mm\,min^{-1}$]	Voltage/[kV]	Current/[μA]
1.032	10.03	12.07
1.032	15.02	18.17
1.032	20.03	24.87
1.032	25.03	31.23
1.032	30.03	38.17
1.526	10.04	13.47
1.526	15.02	20.4
1.526	20.03	27.33
1.526	25.05	35.27
1.526	30.03	43.17
2.018	10.04	16.4
2.018	15.03	24.63
2.018	20.02	33.3
2.018	25.04	45.17
2.018	30.03	62.23
2.512	10.04	18.5
2.512	15.03	27.93
2.512	20.04	45.57
2.512	25.04	71.33
2.512	30.03	118.23

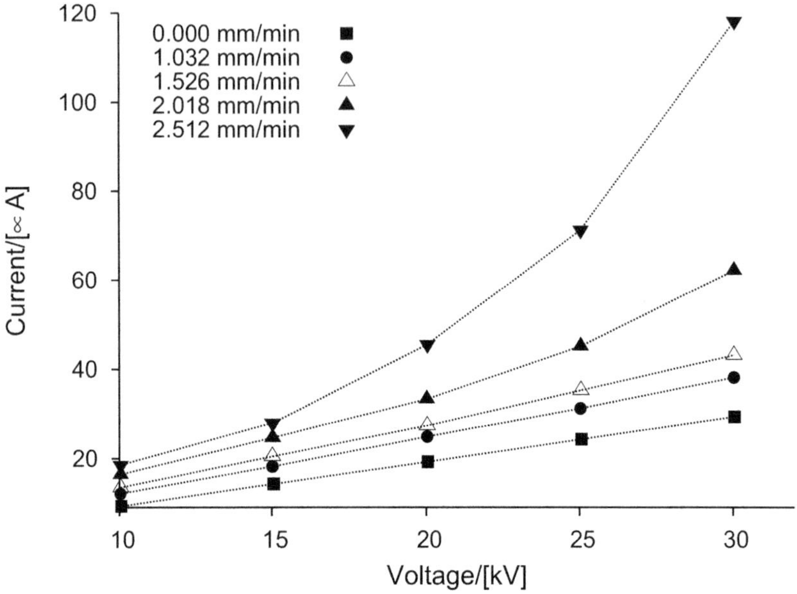

Figure 4.4 Applied voltage viz. leakage current (22).

ally with the increase in the applied voltage. Like under dry conditions, the same tendency of leakage current change also occurred on the insulator under wet conditions (22).

4.3.1.3 Composite Insulators

Silicone elastomer composite insulators and processes for their production are well known. Silicone-rubber injection molding using what is known as solid rubber (high-temperature-crosslinking, HTV, or high consistency rubber, HCR) features injection of silicone rubber of comparatively high viscosity into heated molds.

The process for what are known as hollow insulators is described in the patent EP1091365 (23). The process is currently used for all types of components, including, for example, rod insulators and surge arresters. The cycle times, which are sometimes long, have an adverse effect on the process and result from the requirement that the parts to be sheathed (e.g., fiber-reinforced epoxy-resin rods or corresponding tubes), in particular the metallic add-on parts (fittings) that sometimes protrude from the mold, must likewise be heated to the crosslinking temperature of the rubber. Large components sometimes require machines and apparatuses of considerable size.

Another disadvantage is the presence on the molded component of mold-parting lines, which often require that the moldings be subjected to a subsequent mechanical operation.

A feature common to all of the known processes is that the crosslinking of the electrically insulating material of the exterior sheath of the insulators either occurs spontaneously at room temperature or is initiated thermally at elevated temperature.

The crosslinking of the silicone rubber initiated via the UV irradiation minimizes crosslinking times, and can be used universally for any desired shapes of composite insulators, and is therefore advantageous to the user in relation to total production costs (24).

The method of using UV radiation for the irradiation of the silicone rubber with light should be advantageous for irradiating the silicone rubber volume to be crosslinked in a manner that gives uniformly rapid crosslinking. Irradiation from one direction can sometimes be disadvantageous. In order to achieve uniform irradiation of the silicone rubber, this can be irradiated fully from a plurality of directions.

The silicone rubber can be a mixture composed of two components. The silicone rubber preferably contains (24):

1. A poly(organo siloxane) which contains at least two alkenyl groups per molecule and which has a viscosity of from 0.1 to 500,000 *Pa s* at 25°C,
2. An organosilicon compound comprising at least two SiH functions per molecule, and
3. A platinum-group catalyst activatable by light of from 200 *nm* to 500 *nm*.

Particularly suitable catalysts are cyclopentadienyl complexes of platinum (24).

The silicone rubber is charged to a suitable open casting mold, which is passed along the supportive component to be shielded and which, toward the bottom, has been suitably sealed in such a way that the silicone rubber cannot escape during the charging procedure. Once the charging procedure has been concluded or once a particular fill level has been reached, ultraviolet radiation is used to irradiate the silicone rubber with light, or for intermediate or preliminary irradiation of the silicone rubber with light. The rubber in the casting mold is crosslinked very rapidly during this process (24).

Effect of Fly Ash Filler. Fly ash can be used as a filler material because it is cheap, readily available, and utilizes waste from coal fire power plants (25). The effect of the amount of fly ash as filler in high voltage insulator material of silicone rubber on its electrical properties has been described.

The type of silicone rubber used was RTV 683 (26) and a catalyst in the form of liquid acetoxysilane was also used (27). Various amounts of fly ash from 20% to 50% were loaded into silicone rubber. The electrical properties were measured in the form of relative permittivity, dielectric strength, and surface resistivity. Acetoxysilane is shown in Figure 4.5.

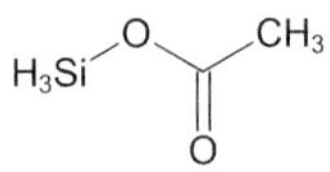

Figure 4.5 Acetoxysilane.

The standard used was the ASTM D150 standard (27). These test methods cover the determination of relative permittivity, dissipation factor, loss index, power factor, phase angle, and loss angle of specimens of solid electrical insulating materials when the standards used are lumped impedances. The frequency range addressed extends from less than 1 *Hz* to several hundred *M Hz*.

The results of the study indicated that the value of the relative permittivity of silicone rubber increases with the increase of fly ash, but it decreases after aging. An increased quantity of fly ash on the silicone rubber made the breakdown voltage and the surface resistivity larger (25).

The voltage breakdown of the test materials was found between 14.1 $kV\ mm^{-1}$ to 16.8 $kV\ mm^{-1}$, where the values are still in the range of breakdown voltage of pure silicon rubber, which is between 9 $kV\ mm^{-1}$ to 39 $kV\ mm^{-1}$ (25).

4.3.2 *Liquid Silicone Rubber Exposed to Acid Fog*

The extensive use of LSR as an insulating material in power transmission and distribution system is due to its excellent electrical properties, high performance under wet and polluted conditions, wide operation temperature range and simple processing (28).

However, in recent years, *acidic attack* is becoming a matter of concern regarding the reliability of silicone rubber outdoor insulation. There are indications that the root cause for the cracking or hydrophobicity loss on some silicone rubber outdoor insulation is a combination of factors, including electrical (corona), environmental (UV, acid rain), mechanical stresses and other effects.

The performance of two types of LSR, energized with DC 1.85 kV voltage (corresponding to a specific creepage distance of 43 $mm\,kV^{-1}$), has been tested under a 1000 h acid fog test. The leakage currents along the surfaces were continuously monitored during the 1000 h test. The hydrophobicity, surface morphology, and hardness of the virgin samples and samples aged after 250 h, 500 h, 750 h, and 1000 h have been evaluated. The experimental results indicated that two types of LSR rubber maintained their hydrophobic surface properties, which was also in agreement with the low leakage currents level throughout the test.

For all the samples, the infrared spectra remained almost unchanged during the whole 1000 h acid fog test, but the hardness increased by around 20%.

All the initial samples and 1000 h-aged samples passed an AC4.5 kV tracking and erosion tests. So, an LSR rubber shows a good performance in the acid fog environment under DC stress (28).

Corona is an unavoidable phenomena in high voltage power transmission system, in spite of suitably designed insulator accessories and transmission line hardware (29,30).

The continuous occurrence of corona can subject the polymeric insulator to a severe degradation. Furthermore, moisture in the air has a positive influence on the corona activity.

The corona performance of the silicone rubber housing material with simultaneous application of coldfog has been evaluated. FTIR showed a higher hydroxylation of the surface of a sample under the moisture application than in the normal condition for both AC and

DC excitation. Also the presence of nitric acid on the treated surface with coldfog application was shown (29).

For polymeric outdoor insulators, the exposure to corona discharges can alter the chemical structure of the basic polymer and it degrades the surface properties. The effect of micro and nano fillers in silicone rubber composites in suppressing such damage has been assessed (31).

Four different types of samples were fabricated: Pristine silicone rubber, micron-sized silica/silicone rubber, micron and nano silica/silicone rubber, and nano silica/silicone rubber composites. The compositions are detailed in Table 4.4.

Table 4.4 Silica particle contents in the composites (31).

Composite	Nano-SiO_2/[%]	Micro-SiO_2/[%]
Pristine silicone rubber	0	0
Micro silica/ silicone rubber composite	0	30
Nano Micro silica/ silicone rubber composite	2.5	27.5
Nano silica/ silicone rubber composite	5	0

The samples were exposed to AC corona using a needle to ground-plane electrode setup. The experimental results are analyzed by phase-resolved partial discharge, hydrophobicity loss-recovery, SEM, surface roughness and FTIR. The results of these measurements indicated that the nano silica/silicone rubber shows a strong resistance to partial discharges and hydrophobicity loss. In the area below the needle tip, higher hydrophobicity loss and higher recovery are observed as compared to the vicinity region. Variations in surface roughness, appearance of crackles, voids, pits, surface splitting into blocky structures and damages to chemical structure of silicone rubber were appreciably retarded in nano silica/silicone rubber as compared to the other samples. It was concluded that addition of nano-sized silica can be an attractive approach to improve the corona resistance of micron-sized silica-filled silicone rubber (31).

An efficient way to fabricate the superhydrophobic SR has been reported whose contact angle reaches 151° and the sliding angle reaches 5° (32). The long-term stability of polymeric superhydrophobic insulating surfaces was examined under electrical and non-electrical stresses. The result shows that in the water aging process, the static contact angle of both SR and superhydrophobic SR can recover, but the dynamic contact angle shows only a little recovery. In the corona aging process, both the static and dynamic contact angle of superhydrophobic SR can almost restore to the original level, but the dynamic contact angle of normal SR increases after restoration. The finding shows that the micro/nano surface structure and the stability of the surface chemical composition are important for the anti-aging performance of the superhydrophobic SR (32).

4.3.3 *Tracking and Erosion Resistance*

An attempt has been made to study the effect of environmental conditions in the service life of polymeric insulators on tracking and erosion resistance (33). Evaluation of tracking and erosion resistance of silicone rubber insulating samples of high temperature vulcanized rubber and LSR were carried out under AC and positive polarity DC voltage.

The samples were treated under normal contaminants as per IEC 60587 and acidic contaminant solution simulating acid rain. The conditions of pollution and contamination severity level were taken into consideration by maintaining the conductivity at approximately 2.5 $mS\ cm^{-1}$ for both solutions. To observe the effect of acidity alone on the samples, the acidity level was varied to understand the material behavior under acidic conditions with AC and DC stress application.

An experimental arrangement was set up based on IEC 60587 and ASTM D2303 standards (34,35). The constant voltage method was employed to find the relative tracking performance of silicone rubber samples for different chemical compositions. Physical chemical analysis involving SEM, energy dispersive X-ray analysis (EDAX) and FTIR was carried out on the so treated samples to understand the surface morphology and the chemical changes.

A comparative analysis was carried out for the acid rain solution and the standard solution of NH_4Cl. Furthermore, analysis of the leakage current performance was carried out under both AC and DC stress (33).

4.3.4 *Color Fading*

The performance of silicone rubber of composite insulators directly affects the safe operation of the transmission lines (36). The aging of silicone rubber is related to not only the complex operating environment, but also the aging resistance ability. In China, a high temperature vulcanized silicone rubber is applied to manufacture the sheds and sheath of composite insulators.

Hardening and color fading are two significant aging phenomena in most cases. The color fading mechanism of high temperature vulcanized silicone rubber has been studied. Also, a method to quantify its color fading degree for field application was developed (36).

In the study, 18 composite insulator specimens were collected and analyzed. Test results showed that there is a positive correlation between the color fading and the microstructural change of the silicone rubber. A quantitative measuring method of the color fading degree of the silicone rubber has been developed. This could be applied in field tests to help field engineers decide the aging level of composite insulators (36).

4.3.5 *Improving Tracking Resistance and Flame Retardancy*

Silicone rubber is increasingly being used as exterior insulator in electrical and electronic applications, such as high voltage transmission lines, bushings and cable terminations, due to its advantages of light weight, high insulation and superior resistance to contamination (37–39)

However, when being employed as exterior insulator in polluted environments, the pollutant and moisture covered on the surface of silicone rubber will result in leakage current and arc discharge. Due to dry band arcing, silicone rubber is prone to tracking and even smoldering combustion, which severely threatens the reliability of the insulation system (40–42)

An effective improvement of the tracking resistance and flame retardancy of silicone rubber is one of the key problems which needs to be resolved in the development of high performance silicone rubber insulating materials (43).

A functional silane containing hindered amine and urethane groups, named PPAS, was synthesized by using 1,2,2,6,6-pentamethyl-4-piperidinol, (3-isocyanatopropyl) triethoxysilane, c.f. Figure 4.6, and allyl alcohol via the nucleophilic substitution and transetherification.

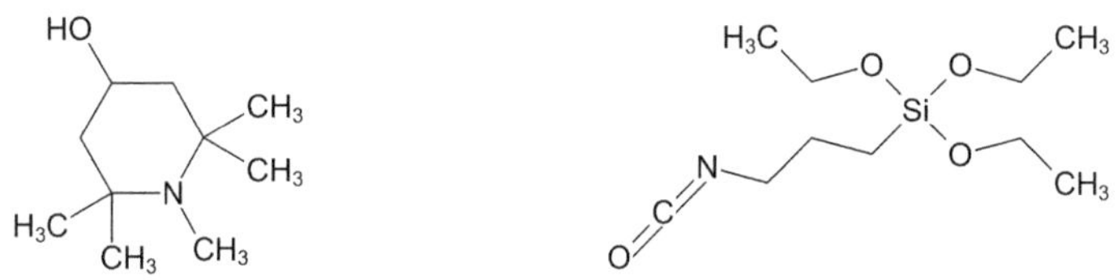

1,2,2,6,6-Pentamethyl-4-piperidinol   (3-Isocyanatopropyl) triethoxysilane

Figure 4.6 Functional silane groups .

The synthesis is shown in Figure 4.7. This synthisis method is similar to that shown in Figure 1.19.

The effects of PPAS on the tracking resistance, flame retardancy and thermal stability of ALSR were investigated. It was found that PPAS could effectively enhance the tracking resistance and flame retardancy of ALSR simultaneously.

An inclined plane test was used to check the electric surface conductivity. Here, the samples were wetted by solutions that contained 0.1% ammonium chloride and 0.2% *i*-octylphenoxypolyethoxyethanol.

The inclined plane test is a standard test method for the evaluation of electrical insulating materials for use under severe ambient conditions (34,35). These test methods quantitatively evaluate the effects on an insulating material resulting from the action of electrical discharges upon a material surface. In these test methods, a conducting liquid contaminant is continuously supplied at an optimum rate to the surface of a test specimen in such a fashion that essentially a continuous electrical discharge can be maintained. The effects are similar to those that may occur in service under the influence of dirt combined with moisture condensed from the atmosphere (34).

When the content of PPAS was only 3.0 *phr*, ALSR/PPAS could pass the 1A4.5 *kV* level in an inclined plane test and had outstand-

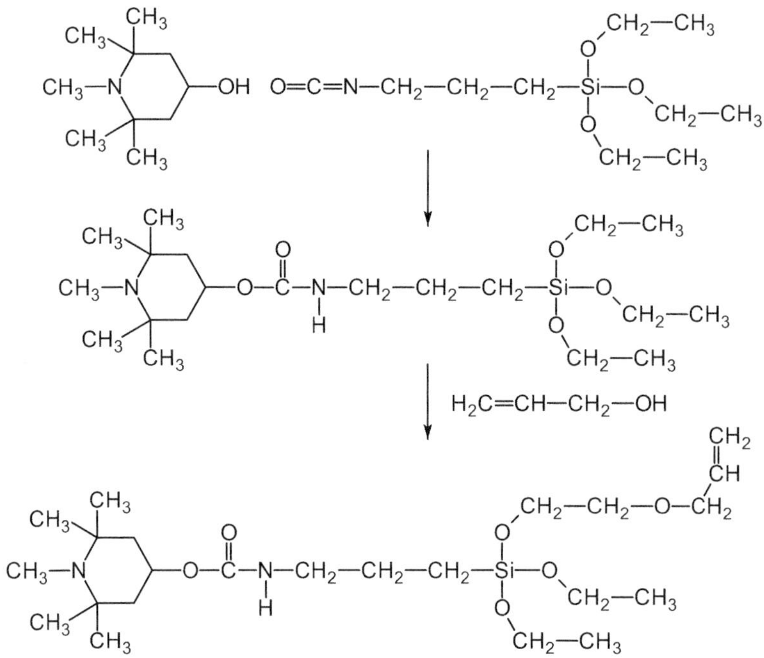

Figure 4.7 Synthesis of a functional silane with hindered amine.

ing electrical erosion resistance, which showed great potential in exterior insulator. In particular, the limiting oxygen index of the ALSR/PPAS was 30.0%, and it had a good self-extinguishing capability. Meanwhile, the risk of fire measured by cone calorimeter was also significantly reduced.

In summary, PPAS could effectively quench the peroxyl radicals and suppress the oxidation degradation of the $Si-CH_3$ groups, eliminate the Si-OH groups and promote the crosslinked reaction during the thermal degradation of ALSR (43).

4.4 Electromagnetic Wave Absorber

The electromagnetic wave absorbing properties of poly(pyrrole)/silicone rubber-filled nanocarbon and carbonyl iron nanocomposites were investigated in the X-band (44).

The poly(pyrrole)/silicone rubber matrix was fabricated by the incorporation of poly(pyrrole) and liquid silicone rubber in a ratio of 10:90. The poly(pyrrole)/silicone rubber serves not only as a matrix but also promotes the flexibility and electromagnetic wave absorption of the nanocomposite sheets.

The chemical, morphology, thermal, wetting, absorption, and mechanical properties of the electromagnetic wave absorber sheets were investigated. An electromagnetic wave absorber sheet with 3% nanocarbon and 37% carbonyl iron and a thickness of 1 *mm*, exhibits a minimum reflection loss of around −13 *dB* at 10.27 *GHz*. The absorber sheet with an elongation of 520% could be qualified as a superior elastic sheet. The specimen represents higher thermal stability compared with pure SR, and the contact angle of the sheet proved that its surface should be hydrophobic (44).

4.5 Suppression of Surface Charge

4.5.1 *Outdoor Insulation Materials*

In modern power systems, the high voltage direct current system is considered the most efficient and economical technology for high-voltage, large-capacity, and long-distance transmission of electrical power (45).

Poly(dimethyl siloxane)-based silicone rubber is widely used in high voltage direct current as an outdoor insulation material because of its low surface energy, low density, and low surface and bulk conductivity over a wide temperature range (46).

Under DC voltages, the sustained electric field makes it easier for charge to accumulate on the insulator surface. The accumulated surface charge has a marked effect on insulation degradation and plays an important role during the development of surface flashover (47,48).

Laser-ablated template and fluoroalkyl silane modified composite coatings have been used to prepare a micro- and nanostructure on silicone rubber surfaces (45).

By testing the surface potential accumulation and dissipation process, the suppression of surface charge on a superhydrophobic silicone rubber was assessed.

The surface physicochemical properties and resistivity were examined. The electron and hole trap distribution were analyzed using isothermal current decay theory.

The peak density of traps associated with the superhydrophobic surfaces were found to be promoted to lower energy levels relative to an unmodified silicone. The surface microstructure increases the trap density and reduces the trap depth. Furthermore, nanocoatings were shown to further reduce surface trap depth.

The improvements to suppress surface charge accumulation are beneficial to a range of electrical industries and help to mitigate a surface flashover on insulation (45).

4.5.2 *Antistatic Compositions*

Since molded articles obtained from a heat curable silicone rubber composition have excellent transparency, they are suitably used for a cover of portable music player, portable game, cellular phone, or controller of a game playing device.

When they are used for the above-mentioned applications, excellent antistatic property is required. This can be achieved by introducing an ionic liquid. However, there arises a problem as the heat resistance and the compression set property of the molded article are lowered. Particularly, when a curing system is an addition reaction, there is a great problem to be solved in which the heat resistance and compression set property, both of which the silicone rubber inherently have, are remarkably deteriorated.

A heat curable addition-reactive type silicone rubber composition has been described, which results in a molded article with excellent antistatic property, and also excellent heat resistance and compression set property. The composition contains (49):

1. A mixture containing a heat curable addition-reactive silicone rubber and a curing agent,
2. An ionic liquid in an amount of 30 to 3000 *ppm* relative to 100 parts by mass of the component 1, wherein the curing agent is not included in the 100 parts by mass,
3. A heterocyclic compound containing nitrogen atoms in an amount of 30 to 3000 *ppm* relative to 100 parts by mass of the component 1, wherein the curing agent is not included in the 100 parts by mass.

As curing catalyst, a platinum-based catalyst can be used, such as platinic acid chloride, platinum olefin complex, platinum vinyl siloxane complex, platinum carbon or a platinum triphenylphosphine complex. As crosslinking agent, a poly(organo siloxane) can be used, in which the number of hydrogen atoms bonding to a silicon atom exceeds at least two on average in one molecule.

From the viewpoint of solubility in water, the ionic liquid is preferably an ionic liquid with a poor water-solubility or water-insolubility. Among the ionic liquids capable of imparting an antistatic property without impairing the excellent properties of the silicone rubber, those having a bis(trifluoromethanesulfonyl)imide anion are most preferable. Such ionic liquids are shown in Table 4.5. Some ionic liquids are also shown in Figure 4.8.

Preferred heterocyclic compounds are 5,5′-bi-1H-tetrazole-2-ammonium and 5,5′-bi-1H-tetrazole-piperazine (49).

The synthesis of ionic liquids have been detailed (50). For example, 3-methyl-1-trimethoxysilylpropylpyridinium bis(trifluoromethanesulfonyl)imide has been prepared as follows:

Preparation 4–3: Under N_2 atmosphere, 60 *mmol* of 3-methylpyridine, c.f. Figure 4.9, (a reagent manufactured by Tokyo Chemical Industry Co., Ltd.) and 55 mmol of 3-chloropropyltrimethoxysilane (a reagent manufactured by Tokyo Chemical Industry Co., Ltd.) were mixed and reacted at 90°C for 72 *h*. After cooling the reaction mixture, the precipitated

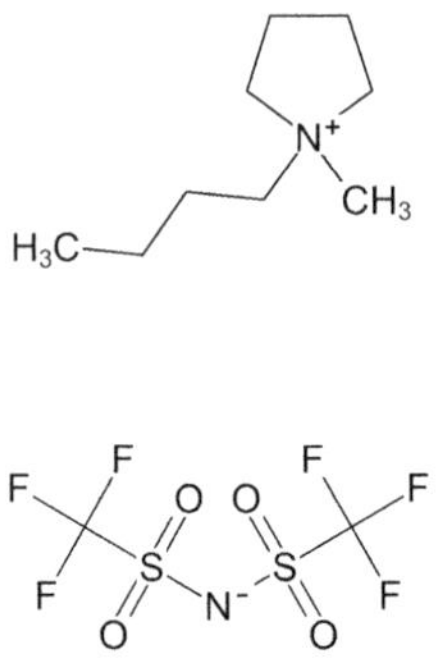

1-Butyl-1-methylpyrrolidinium
bis(trifluoromethanesulfonyl)imide

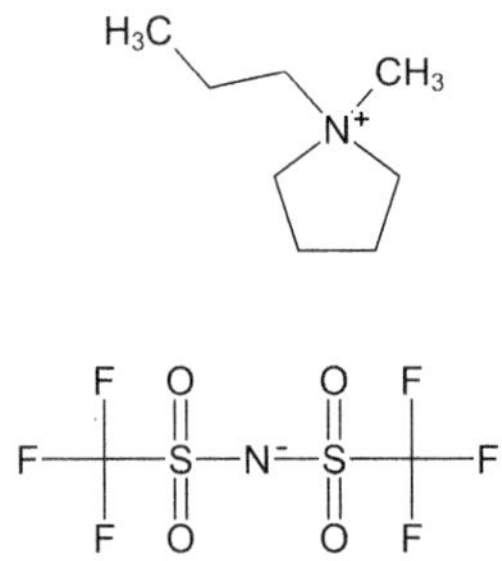

1-Methyl-1-propylpyrrolidinium
bis(trifluoromethanesulfonyl)imide

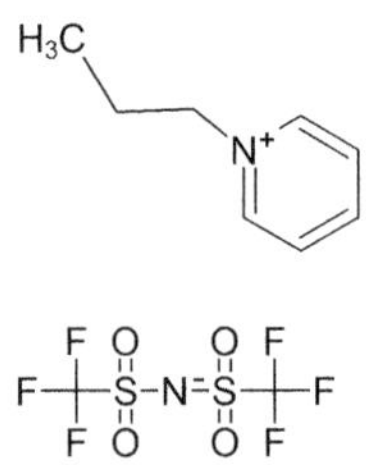

3-Methyl-1-trimethoxysilylpropylpyridinium
bis(trifluoromethanesulfonyl)imide

Figure 4.8 Ionic liquids (49).

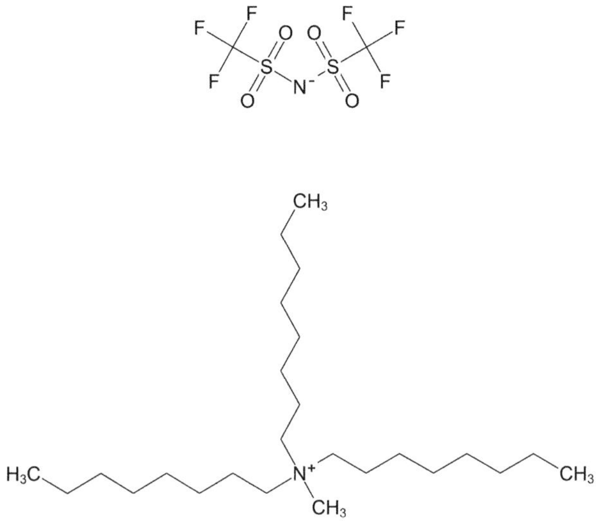

Methyltrioctylammonium
bis(trifluoromethanesulfonyl)imide

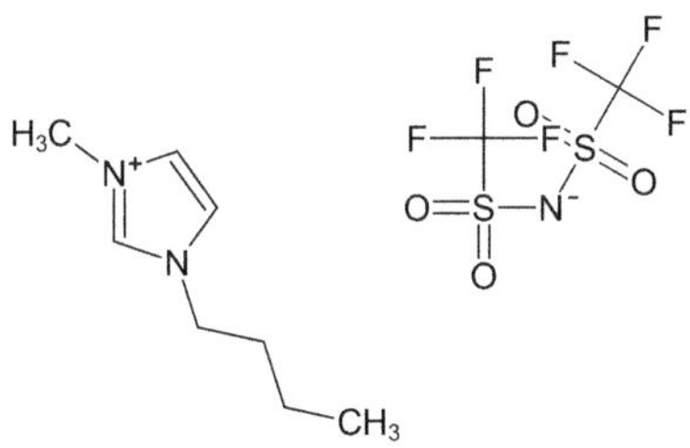

1-Butyl-3-methylimidazolium
bis(trifluoromethanesulfonyl)imide

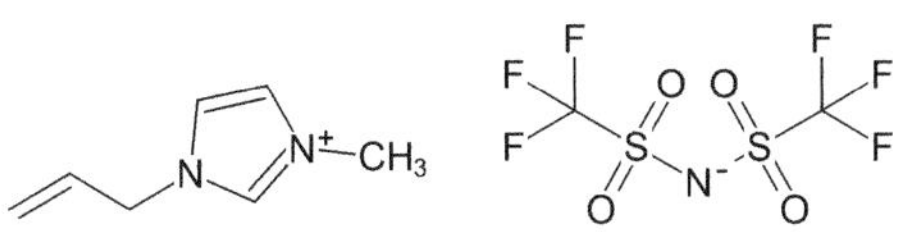

3-Methyl-1-trimethoxysilylpropylpyridinium
bis(trifluoromethanesulfonyl)imide

Figure 4.8 (cont.) Ionic liquids (49).

Table 4.5 Ionic liquids (49).

Compound
1-Butyl-1-methylpyrrolidinium bis(trifluoromethanesulfonyl)imide
1-Methyl-1-propylpyrrolidinium bis(trifluoromethanesulfonyl)imide
3-Methyl-1-propylpyridinium bis(trifluoromethanesulfonyl)imide
N-Butyl-3-methylpyridinium bis(trifluoromethanesulfonyl)imide
1-Methyl-1-propylpyridinium bis(trifluoromethanesulfonyl)imide
Diallyldimethylammonium bis(trifluoromethanesulfonyl)imide
Methyltrioctylammonium bis(trifluoromethanesulfonyl)imide
1-Butyl-3-methylimidazolium bis(trifluoromethanesulfonyl)imide
1,2-Dimethyl-3-propylimidazolium bis(trifluoromethanesulfonyl)imide
1-Ethyl-3-methylimidazolium bis(trifluoromethanesulfonyl)imide
1-Vinylimidazolium bis(trifluoromethanesulfonyl)imide
1-Allylimidazolium bis(trifluoromethanesulfonyl)imide
1-Allyl-3-methylimidazolium bis(trifluoromethanesulfonyl)imide

solid was washed twice with ethyl acetate, and then ethyl acetate was removed under reduced pressure to produce 53 *mmol* of a compound of 3-methyl-1-trimethoxysilylpropylpyridinium chloride. The resulting compound was dissolved in acetone, and 53 *mmol* of bis(trifluoromethanesulfonyl)imide lithium (a reagent manufactured by Tokyo Chemical Industry Co., Ltd.) was added thereto, followed by stirring for 24 *h* at room temperature. The solvent was removed under reduced pressure, and the precipitated lithium chloride was filtered to produce 40 *mmol* of 3-methyl-1-trimethoxysilylpropylpyridinium bis(trifluoromethanesulfonyl)imide

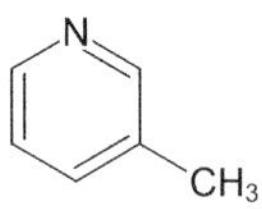

Figure 4.9 3-Methylpyridine.

4.6 Heat Dissipation Devices

In electronic devices, e.g., notebook personal computers, the amount of heat generated from internal components has increased in accordance with improvements in the performance of the devices (51). In the case where such an electronic device is used, when the outside air temperature is high or the condition of the installation location of the device is bad, the heat is not dissipated sufficiently and the operation of the device may become unstable.

4.6.1 *Liquid-Encapsulation Heat Dissipation Member*

A liquid-encapsulation heat dissipation member has been described, which is prepared by encapsulating a thermally conductive fluid in a closed container (51). This dissipates the heat generated from an electronic device placed on the closed container, wherein the closed container includes an elastic portion composed of an elastomer serving as a surface to come into contact with the electronic device. The thermally conductive fluid contains a thermally conductive powder and has a viscosity of 200,000 $mPa\,s$ to 3,000,000 $mPa\,s$.

The heat dissipation sheet was obtained by mixing 180 parts by weight of aluminum hydroxide (crushed, particle diameter of 50 μm) and 180 parts by weight of aluminum hydroxide (crushed, particle diameter of 8 μm) into 100 parts by weight of liquid silicone rubber and, thereafter, performing heat curing. The outer shape was 110 $mm \times 75\ mm \times 5.0\ mm$. The thermal conductivity of the heat dissipation sheet was 2.0 $W\,m^{-1}K^{-1}$. The thickness of the aluminum plate was 1 mm.

The liquid-encapsulation heat dissipation member has a durability, so it is not damaged by contact such as repetitive placement of an electronic device. It has flexibility so as to follow the outside surface of even a small electronic device placed thereon, and it shows a high cooling efficiency (51).

4.6.2 *Loop Heat Pipe*

A compact and highly efficient loop heat pipe can be used as a cooling device for suppressing a temperature rise due to heat from a heat-generating part in electronic equipment (52). A schematic view of the device is shown in Figure 4.10.

Here, a working fluid is sealed in the interior of the loop heat pipe. The loop heat pipe includes an evaporator that absorbs heat from a heat-generating part and evaporates the working fluid from a liquid

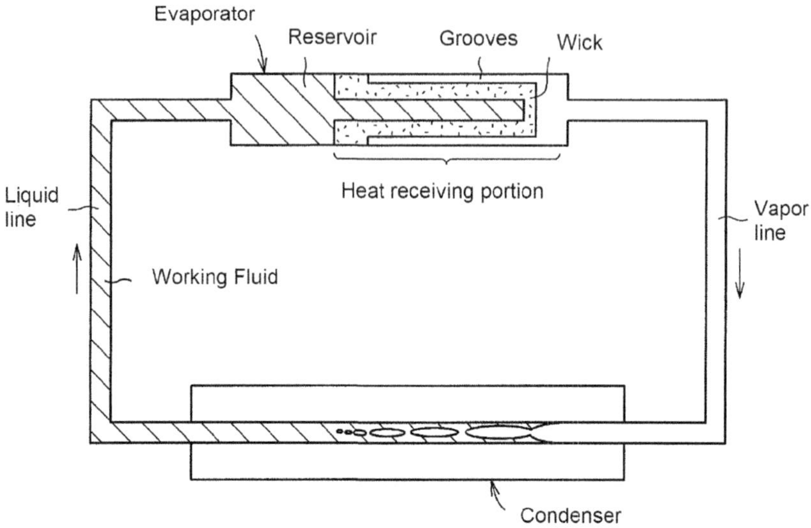

Figure 4.10 Loop heat pipe (schematic) (52).

phase to a gas phase, a condenser that condenses a gas-phase working fluid leading from the evaporator into the liquid phase, a vapor line that allows the gas-phase working fluid to flow from the evaporator to the condenser, and a liquid line that allows a liquid-phase working fluid to flow from the condenser to the evaporator.

The evaporator is a cylindrical member formed of a metal having good thermal conductivity, such as copper or a copper alloy. The evaporator includes a heat receiving portion in which the wick is accommodated and a reservoir that stores the liquid-phase working fluid. The heat receiving portion is coupled to an end of the vapor line, and the reservoir is coupled to an end of the liquid line. The other end of each vapor line and liquid line is coupled to the condenser. The condenser is a stainless steel pipe having a large number of thin plate-shaped fins made of aluminum on its outer peripheral surface.

The wick is a hollow member made of a porous material. It is open on the vapor line side and closed on the reservoir side. A plurality of grooves extending in the longitudinal direction are disposed on the outer peripheral surface of the wick over an area from an end of the wick in the vapor line side to a short distance from another end of the wick of the reservoir side.

A foamed silicone rubber or a foamed urethane rubber is used as the porous rubber for the wick (52). To produce the foamed silicone rubber by the water foaming method, a catalyst, a surfactant, and a crosslinking agent are added to a commercially available two-liquid type liquid silicone rubber and mixed. If necessary, additives, fillers, dispersants and the like are mixed with water as a solvent, if necessary, mixed with alcohol. The mixture is stirred together with a mixed solution having the same viscosity as the liquid silicone rubber to prepare an emulsion composition.

The compounding ratio of the liquid silicone rubber and the mixed solution is adjusted according to the desired porosity. As the particulate moisture in the emulsion evaporates to form a cell, when the blending ratio of the liquid silicone rubber and the mixed solution is 1:1, a porous body with a porosity of 50% is obtained.

For the preparation of the emulsion, a homogenizer and, if necessary, a stirring machine for ultrasonic treatment is used. Various stirring conditions, such as stirring means, stirring time and stirring speed (for example, 300 *rpm* to 1500 *rpm*), are adjusted in order to obtain a cell distribution that satisfies the above conditions.

After stirring, the so prepared emulsion composition is poured into a mold, and primary heating is carried out to harden the silicone rubber without evaporating water in the emulsion composition and mold the emulsion composition into a predetermined shape (a primary heating and molding process). The primary heating is carried out preferably at a heating temperature of 90°C to 110°C and a heating time of 60 *min* to 90 *min*.

Then, a secondary heating is done to remove moisture from the porous body after primary heating. The secondary heating is preferably carried out at a heating temperature of 200°C to 250°C and the heating time is 3 *h* to 5 *h*. In the process of secondary heating, moisture is removed from the porous body, and the final curing of the silicone rubber is completed.

The properties of the wick made of water foamed silicone rubber were tested. The following results have been reported (52):

A wick made of water foamed silicone rubber can effectively cool the heat-generating part when the wick has an average pore size of 50 μm or less, the porosity of 20% or more and 80% or less, and the continuous cell ratio of 25% or more and 100% or less. However, in the sample having the average pore diameter of more than 10 μm, the maximum water head difference at which the loop heat pipe can drive is lower than that of other samples having the average pore diameter of 10 μm or less. Even when the porosity and the continuous cell ratio are almost the same, as the average pore size increases, the maximum water head difference becomes smaller. It is thought that this occurs because the capillary force is reduced by the large average pore size. Therefore, the test results show that an average pore size of 10 μm or less is preferable in order for the capillary force of the wick to move the working fluid and drive the loop heat pipe more effectively (52).

4.7 Optical Fiber Sensor

A silicone rubber-coated Mach-Zehnder interferometer has been proposed and used for temperature measurement (53).

A Mach-Zehnder interferometer is a device used to determine the relative phase shift variations between two collimated beams derived by splitting light from a single source (54). The interferometer can be used to measure phase shifts between the two beams caused by a sample or a change in length of one of the paths. The apparatus is named after the physicists Ludwig Mach, the son of Ernst Mach, and Ludwig Zehnder.

The Mach-Zehnder interferometer was fabricated by splicing a single mode fiber between a short section of no-core fiber and the ultra-abrupt taper region (53).

The sensing length of the Mach-Zehnder interferometer was coated with a liquid silicone rubber to enhance the temperature sensitivity. Here, no-core fiber was used to excite the higher order cladding mode, the ultra-abrupt taper region acts as an optical fiber coupler, and the silicone rubber coating on sensing length was used as solid cladding material instead of liquid.

The enhancement of the sensitivity of a device occurs due to the high refractive index of 1.42 and the thermo-optic coefficient ($-1.4\times10^{-4}/°C$) of silicone rubber as compared to liquid cladding temperature sensors.

The experiment was performed for both coated and uncoated Mach-Zehnder interferometers and the results were compared. The

Mach-Zehnder interferometer exhibits a high temperature sensitivity of 253.75 *pm*/°C and 121.26 *pm*/°C for coated and uncoated sensing probes, respectively, in the temperature range from 30°C to 75°C (53).

4.8 Optical Semiconductor Device

Silicones can form cured products with excellent weather resistance, heat resistance, and similar properties. Therefore, curable silicone compositions are used as sealing agents, protective agents, or coating agents for semiconductor elements in optical semiconductor devices such as photocouplers, light-emitting diodes, or solid-state image sensing devices (55).

Previously proposed curable silicon compositions that were used as sealing resins for semiconductor elements showed poor thermal shock resistance. Thus, there have been problems such as the ready occurrence of cracks in the resin due to thermal shock, and particularly failures such as non-lighting when such a semiconductor element is a light-emitting diode.

The optical semiconductor device is shown in Figure 4.11 in a cross-sectional view of a stand-alone surface-mounted light-emitting diode (LED) device that is an example of an optical semiconductor device. The LED device is produced by using an adhesive for die bonding of an LED chip on a die pad within a poly(phthalamide) resin housing, performing wire bonding by a bonding wire between the LED chip and an inner lead, and using a sealing material to seal the LED chip, the inner lead, and the bonding wire together with the inner wall of the resin housing. In the optical semiconductor device, the curable silicone composition is used as the composition for forming the adhesive and/or sealing material.

A silicon composition with good thermal shock resistance has been described (55). The composition contains:

1. An organopolysiloxane having at least two aliphatic unsaturated hydrocarbon groups per molecule,
2. An organohydrogenpolysiloxane having at least two silicon-bonded hydrogen atoms per molecule,

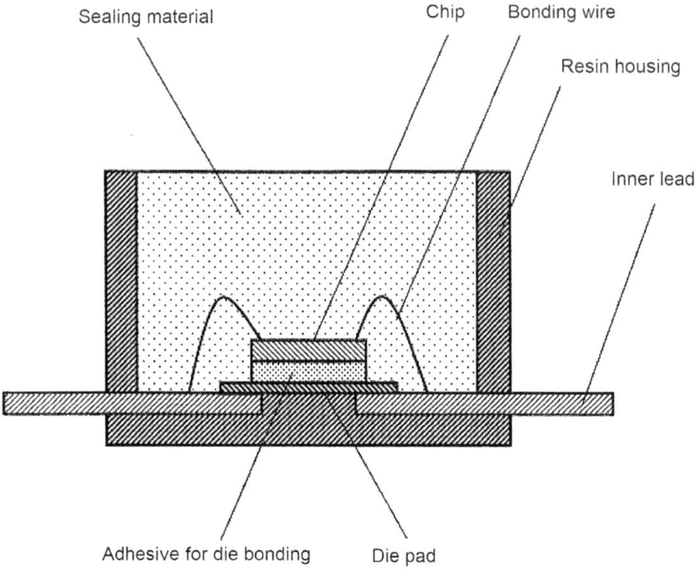

Figure 4.11 Optical semiconductor device (55).

3. At least one type of compound selected from the group consisting of triazole-based compounds with the exception of non-substituted benzotriazole, and
4. A hydrosilylation reaction catalyst.

The triazole compounds are preferably *N,N*-bis(2-ethylhexyl)-[(1,2,4-triazol-1-yl)methyl]amine, tolyl triazole, carboxy benzotriazole, and nitrobenzotriazole, c.f. Figure 4.12.

4.9 Light-Emitting Devices

4.9.1 Composition for a Light-Emitting Diode

A simple approach has been developed to get a multiscale network of heat conducting by filling spherical alumina oxide and graphene nanoplatelets into SR (56).

This unique structure effectively minimizes the thermal contact resistance between fillers and matrix. The physical properties of the composites were characterized by thermal conductivity, density,

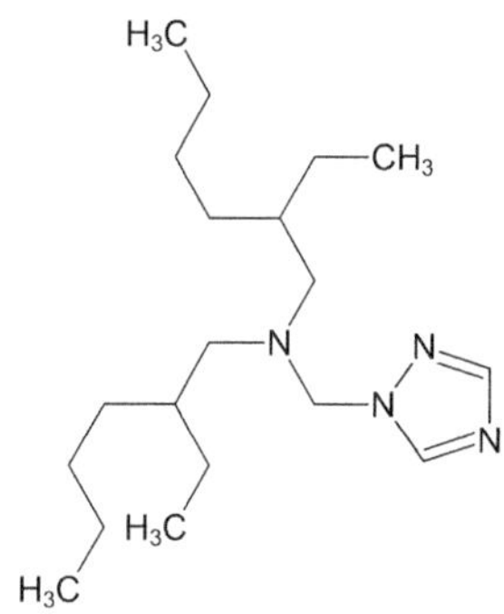

N,N-Bis(2-ethylhexyl)-[(1,2,4-triazol-1-yl)methyl]amine

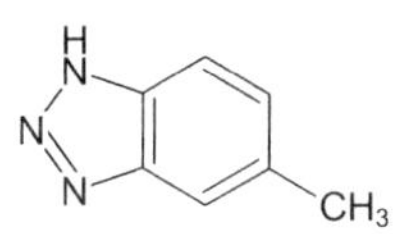

Tolyl triazole

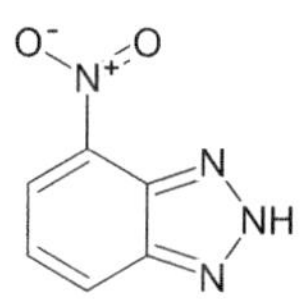

4-Nitro-2H-benzotriazole

Figure 4.12 Triazole compounds.

and tensile strength. A high thermal conductivity of 3.37 $W m^{-1} K^{-1}$ has been achieved, which is 47.1% higher than the single filler at the same loading.

The wet functionalization of graphene and its applications in rubber composites has been described in a monograph (57).

A strong and obvious synergistic effect has been observed as S-Al2O3 and graphene nanoplatelets filled into silicone rubber matrix. It is interesting that the composites with graphene nanoplatelets have the lower density of 2.62 $g cm^{-}3$, reduced by 6%, and the superior tensile performance, compared to a silicone rubber composite with neat alumina oxide. The composites should have potential applications in heat dissipation of a light-emitting diode (56).

4.9.2 *Encapsulating Materials*

Silicone rubber has attracted considerable attention from LED manufacturers, because of its superior insulation, excellent weather and UV resistance and outstanding thermal shock resistance (58).

The technical principle used in silicone encapsulant synthesis has been assessed, which shows how each raw material works.

Also, some product design experiences were illustrated with an example of designing a silicone encapsulant with excellent sulfur corrosion resistance. The future development trend of silicone encapsulants has also been proposed, expecting that more researchers can draw their inspiration from this article (58).

In recent years, a cured product of a silicone resin composition has been proposed as an encapsulating material for a semiconductor light-emitting element (59). However, a cured product of a silicone resin composition has a high gas permeability and a low barrier property against hydrogen sulfide gas in air.

For this reason, there has been a problem in that, when encapsulating is made by using a cured product of a silicone resin composition, a silver film which is a back-surface reflection plate of the encapsulated semiconductor light-emitting element is corroded by hydrogen sulfide in air, thereby leading to a decrease in the brightness of the semiconductor light-emitting element.

As a silicone resin composition that solves the aforementioned problem, a curable silicone resin composition comprising a silicone resin was proposed in which the refractive index of the silicone resin composition after curing is 1.50 to 1.55 and a silicon oxide filler having an average particle size of 1 to 10 μm and being homogeneously dispersed at a concentration of 1% to 30% in the silicone resin (59).

4.9.3 Waterproof LED Lamp

A conventional locking structure for a waterproof LED lamp contains multiple threaded orifices formed on a casing, a plurality of screws or screw bolts, and multiple nuts, such that the waterproof LED lamp is fixed by screwing the multiple threaded orifices, the plurality of screws or screw bolts, and the multiple nuts together. However, water flows into the waterproof LED lamp easily at a high water pressure (60).

A bolt locking structure for a waterproof LED lamp has been developed, which locks the first casing and the second casing securely (60).

The first casing includes multiple first protrusions and multiple first recesses which are arranged on each of four peripheral walls thereof respectively, wherein any two adjacent devices of the multiple first protrusions are separated by each of the multiple first recesses, and any two adjacent devices of the multiple first recesses are separated by each of the multiple first protrusions.

The second casing includes multiple second protrusions and multiple second recesses which are arranged on each of four peripheral walls thereof respectively, wherein any two adjacent devices of the multiple second protrusions are separated by each of the multiple second recesses, and any two adjacent devices of the multiple second recesses are separated by each of the multiple second protrusions.

A method of assembling the bolt locking structure contains steps of (60):

1. Retaining multiple first protrusions of a first casing with multiple second recesses of a second casing, and retaining multiple first recesses of the first casing with multiple second protrusions of the second casing,
2. Applying liquid silicone rubber on each of four connection bolts, and inserting each connection bolt through multiple through holes on each of four peripheral walls of each of the first casing and the second casing respectively, and

3. Solidifying the liquid silicone rubber so that each connection bolt is adhered in the multiple through holes.

4.9.4 *High Precision Optics*

An injection molding process of combined primary optics and optical heads made of LSR was investigated (61). Due to the material-specific advantages of highly transparent LSR, especially regarding the excellent high temperature resistance and the great freedom in designing optics that can meet both the requirements for the encapsulation of LED chips as well as the realization of an optical function.

An integrated injection mold concept was realized in order to reduce the mounting steps, where an LED circuit board is inserted and directly overmolded with LSR. Processing studies have shown that a reproducible replication of LSR optics with good optical properties and molding accuracy is possible. Moreover, for use in an automotive application, long-term properties are crucial. Therefore, a temperature cycling test between −40°C and 125°C for 1,000 cycles as well as a continuous operation of the LED with permitted maximum current for 2,000 *h* were carried out.

In the conventional design of a high-beam module, the semiconductor chip of the LED is first encapsulated with a non-injection-moldable casting compound. The encapsulation used to protect the LED against environmental influences and mechanical stresses, simultaneously represents the first optical element and is therefore also referred to as primary optics.

In order to increase the economical attractiveness, the LSR containing optics is injected directly onto the LED board. Therefore, the LED board is inserted into the opened mold and kept in the desired position by centering pins. Afterwards, the liquid silicone rubber is injected through an opening in the back of the board with a cold runner system. In this way, cost- and time-consuming mounting steps can be reduced (61).

Actually, the molding temperature has a strong influence on the crosslinking behavior of the LSR. In addition, the molding temperature determines the thermal stresses on the chip during injection molding. In order to exclude an unnecessarily long exposure of the chip to heat and to design the process as economically as possible, the crosslinking time was adjusted to the mold temperature. On the other hand, the injection flow rate influences the shear stress on the chip during the injection phase.

It was shown that the composition and the LED fulfills long-term requirements with regard to temperature and maximum current. After both long-term tests, no significant color shift was measured. However, a color shift to blue light is seen directly after overmolding (61).

4.10 Capacitance Sensors

The dielectric properties of $CaCu_3Ti_4O_{12}$-silicone rubber composites have been evaluated (62). Composites with different $CaCu_3Ti_4O_{12}$ mass fractions have been prepared. SEM images showed that inorganic particles were distributed evenly in the silicone rubber matrix.

The experimental results revealed that the composites possess excellent mechanical and dielectric properties. The dielectric constant of the composites has been simulated in light of two different theoretical models which can adequately predict the experimental values for the composites. Except when the mass fraction of $CaCu_3Ti_4O_{12}$ increases beyond 30%, the experimental values begin to deviate slightly downward from the predicted values.

Both the dielectric constant and loss values increase with the increase of uniaxial pressure applied on the composites. For the composite with a $CaCu_3Ti_4O_{12}$ mass fraction of 50% under the pressure of 4 *MPa*, its dielectric constant value has been improved by about three times, but its loss value has been raised by less than twice that of the composites without pressure. Therefore, the dependence of dielectric constant on pressure makes it a promising material in practical applications such as capacitance sensors and other electronic devices (62).

4.11 Dielectric Elastomer Transducers

Silicone elastomers are promising materials for dielectric elastomer transducers due to their superior properties such as high efficiency, reliability, and fast response times (63). Dielectric elastomer transducers consist of thin elastomer films sandwiched between compliant electrodes, and they constitute an interesting class of transducer due to their inherent light weight and potentially large strains.

The current state of silicone elastomers for dielectric elastomer transducers has been reviewed and critically discussed, including commercial elastomers, composites, polymer blends, grafted elastomers and complex network structures (63).

For the field to progress towards industrial implementation, a leap in material development is required, specifically targeting longer lifetime and higher energy densities to provide more efficient transduction at lower driving voltages.

Electroactive polymers consist of soft polymeric materials with relatively large energy densities, relatively fast response times, and the ability to undergo large deformations when stimulated by electrical fields. Electroactive polymers can be divided into two main groups: ionic and electronic. Ionic electroactive polymers respond to an electrical field through the diffusion of ions in a material. In addition, they generally require low driving voltages of 1 V to 2 V, but require an electrolyte, which means that encapsulating the aqueous environment is necessary. Furthermore, there is a need to maintain wetness, and it is difficult to maintain DC-induced displacements.

However, recently it has been proven that it is possible to create semi-interpenetrating networks from polyionic liquids, which eliminates the need for an aqueous environment and thereby opens up the possibility of a wide range of applications where large forces are not needed.

Silicone elastomers are one of the most used materials for dielectric elastomer transducers due to their high efficiency, reliability, and fast response times (64,65).

Frequently used LSRs for dielectric elastomer transducers are Elastosil LR 3043/30 and 3043/50 from Wacker and NuSil CF19-2186. NuSil CF192-2186 was shown to be an ideal candidate for dielectric elastomers, as it possessed the highest measured energy density and strain, as well as a better coupling efficiency, low creep, and excellent tolerance to temperature and humidity (66,67). More details can be found elsewhere (68,69).

Silicone elastomers may be partly self-healing upon electrical breakdown, but if they are torn, the film cannot, with current materials, be restored in a simple way (63).

In summary, commercially available elastomers that have been tested as dielectric elastomer transducers have shown great promise with respect to both actuation and reliability, but they still suffer from having been originally designed for other applications. The dielectric permittivities of commercial silicone elastomers still remain too low for most practical dielectric elastomer transducer applications due to high driving voltages.

The simplest approach to increase dielectric permittivity is to add high dielectric permittivity fillers. The advantage of this approach is in particular the easy preparation, as well as the large variety of fillers available from a large number of suppliers.

Methods, such as roll milling or speed mixing, seem to be the most promising approaches for an industrial application to be implemented and up-scaled for industrial production. Also, the stability needs to be fully tested for the most promising composite systems in order to determine the optimal level of loading (63).

4.12 Solar Cells

In a typical photovoltaic solar cell module, crystalline or amorphous semiconductor solar cells are interconnected, mechanically supported, and protected against environmental degradation by integrating the cells into a laminated structure (70). The modules usually contain a transparent front panel or substrate, preferably made from glass and a rear *backskin* or superstrate, usually a plastic or a metal sheet and sometimes glass. The semiconductor solar cells are usually partly or completely encapsulated by an encapsulating material, and interconnected solar cells are located between the front panel and the backskin.

Originally, silicone encapsulants were found to be too expensive or too complicated for broad use in terrestrial modules. Organic polymers became more important, with ethylene-vinyl acetate (EVA) copolymer considered the best encapsulant for modules with crystalline solar cells. However, EVA tends to discolor with time under the influence of sunlight, a process which reduces the efficiency of the module. Another disadvantage is the need for vacuum laminators for the production of EVA encapsulated modules.

To avoid this time-consuming and complicated process, a process would be desirable for an economical module production. Silicone encapsulants have again become a matter of interest, especially the addition-cured materials, because of the absence of curing by-products (71).

One of the most important conditions for the applicability of silicones is the adhesion to substrates, which are in contact with the encapsulant. Addition-cure silicone materials usually do not adhere well to most substrates (70).

It has been found that a silicone encapsulant for photovoltaic modules that satisfies the above-mentioned needs requires a special selection of the Si–H crosslinking agents.

Curable poly(organo siloxane) compositions for the encapsulation of photovoltaic modules have been described (70).

The curable compositions are liquid at room temperature (25°C) in the uncured state, and cure to elastomers and develop permanent adhesion, in particular, to adjacent glass and plastic substrates, with firm adhesion particularly after long-term outside weathering or after corresponding accelerated aging tests.

A filler batch can be produced as follows (70):

Preparation 4–4: First, 22.5 *kg* of a vinyl-terminated linear poly(dimethyl siloxane) with a viscosity of 10 *Pa s* at 25°C was placed in a planetary mixer and mixed with 2.8 *kg* of hexamethyldisilazane and 0.9 *kg* of water. Then, 12.0 *kg* of fumed silica with a BET surface of 300 $m^2 g^{-1}$ was gradually added and mixed in until a stiff mixture was obtained. This mixture was stirred and heated to reflux for 30 *min*. The volatiles were distilled off and subsequently vacuum was pulled for 30 *min*. The mixture was diluted with 7.8 *kg* of the above poly(dimethyl siloxane). When the resulting filler batch was used for the preparation of the following encapsulants it was calculated as having 28.3% of silica and 71.7% of vinyl-terminated polymer with a viscosity of 10 *Pa s*.

As crosslinking agents, poly(organo hydrogen siloxane)s can be used having at least seven Si atoms, wherein the molar ratio of the Si–H groups to all the Si atoms is more than 0.55, preferably 0.6, with a particular preference of 0.7 to 0.95. Encapsulants with different crosslinking agents can be produced as follows (70):

Preparation 4–5: First, 52 *g* of the filler batch from preparation 4–4(F 1) was mixed with the different crosslinking components in %, 0.12 *g* of a platinum solution of the Karstedt type with 1% of platinum, 10.5 μl of inhibitor ethinylcyclohexanol. Then the mixture was completed up to 100 *g* with ca. 46.5–47.5 *g* of a vinyl-terminated linear poly(dimethyl siloxane) having a viscosity of 10 *Pa s*.

Also, a light-activatable metal catalyst can be used, such as trimethyl(methylcyclopentadienyl)-platium(IV), dissolved in a vinyl-terminated poly(dimethyl siloxane) (70). Light-activatable compositions are particularly preferred because they do not require heat curing at elevated temperatures, thereby reducing the thermal stress to the solar cell modules, in particular to the photovoltaic modules. Furthermore, these compositions can be cured *on command*, without potting times or shelf life times.

4.12.1 *Foamed Sealing Materials*

Foamed bodies can be used as sealing materials in solar cell-related fields (72). Such a sealing material used for a solar cell is disposed between the peripheral end portion of a panel and a support frame material and prevents the entry of water and the like into the panel, for example, when the peripheral end portion of the solar cell panel is fixed to the support frame member.

Such sealing materials can be obtained by foaming rubbers, such as EPDMs, with foaming agents such as azodicarbonamide, acrylic foamed bodies.

However, it is desired that the sealing materials that are used for solar cells exhibit high shock absorbency and sealing properties even if the thickness is small. In addition, since solar cells are installed and used outdoors for a long period, it is desired that the sealing materials have high cold and heat resistance and light resistance so that performance is maintained even if temperature changes due to a difference in temperature between day and night or between the four seasons occur.

A silicone resin is widely known as a material with high cold and heat resistance and high light resistance. However, it is difficult to manufacture a foamed body composed of a silicone resin that has high shock absorbency and sealing properties even if the thickness is small (72).

It has been found that by dispersing a plurality of particles, each having a cavity portion therein, in a silicone resin to make the cavity portions of the cells of a foamed body, and also by keeping a void between the particles as a cavity without filling the void with a silicone resin, a silicone resin foamed body having a high expansion ratio can be manufactured (72). It has also been found that a silicone resin foamed body itself and a laminated body formed by further laminating a film on the foamed body have good shock absorbency, sealing properties, cold and heat resistance, and light resistance even if the thickness is small, and are useful for solar cells.

A method for manufacturing a silicone resin foamed body is done by the following steps (72):

1. A plurality of unfoamed foamable particles, such as expandable microcapsules, are foamed to obtain particles each having a cavity portion therein. It is preferable that the unfoamed foamable particles are added to a base resin of a silicone resin composition and then the resultant mixture is heated, so that the unfoamed foamable particles are expanded.
2. The particles that have been mixed with the base resin in Step 1 are then mixed with the remaining silicone resin composition to prepare a resin-particle mixture.
3. The so obtained resin-particle mixture is disposed on a film, so that the thickness becomes uniform.
4. The resin-particle mixture that has been disposed on the film is now heated to cure the silicone resin composition, resulting in the silicone resin foamed body.

When the volume of the cavity portion could be readily increased, a high expansion ratio could be obtained, although the thickness was small, and a foamed body with good 20% and 50% compressive stresses and also excellent shock absorbency and sealing properties could be obtained (72).

4.13 Portable Electronic Devices

As portable electronic devices become more compact, and the number of functions performed by a given device increase, it has become a significant challenge to design a user interface that allows users to easily interact with a multifunction device (73). This challenge is particularly significant for handheld portable devices, which have much smaller screens than desktop or laptop computers. This situation is unfortunate because the user interface is the gateway through which users receive not only content but also responses to user actions or behaviors, including user attempts to access a device's features, tools, and functions.

Some portable communication devices, such as mobile telephones, have resorted to adding more push buttons, increasing the density of push buttons, overloading the functions of push buttons, or using complex menu systems to allow a user to access, store and manipulate data. These conventional user interfaces often result in complicated key sequences and menu hierarchies that must be memorized by the user. Thus, there is a need for the development of improved wearable devices with sensors.

A wearable device has been described that is at least partially made of a silicone rubber. Furthermore, a support member is at least partially positioned in the wearable device structure (73).

4.14 Cable Accessories

4.14.1 Water Diffusion Model

Liquid silicone rubber (LSR) is a vital component of cable accessories and its dampness is one of the most important reasons for cable faults (74). Therefore, the investigation of the moisture diffusion property and moisture-induced electrical performances of LSR is of great significance.

A diffusion model of water molecules in LSR and the influence of temperature on the diffusion process of water molecules were studied. Furthermore, the effect of water absorption on the electrical properties of liquid silicone rubber was investigated.

The results indicated that the diffusion process conforms to the Langmuir diffusion model, rather than to the classical Fick diffusion model. This means that water exists in LSR in a free state and in a bounded state (74). Also, the temperature affects the diffusion process of the water molecules and the dissolution process of low-molecular-weight siloxane and filler particles to some extent, for which the saturated water absorption rate increases first and then decreases with increasing temperature.

In addition, the water absorbed by LSR decreased the volume resistivity and the AC breakdown strength of LSR, and increased the dielectric constant and loss tangent of LSR (74).

4.14.2 *Cold Shrink Splices*

Cold shrink splices are used to achieve a tight fit around cables and bundle cables together for protection against environmental factors (75). A conventional cold shrink product typically comprises a flexible tube of an elastomer, which is held in expanded condition on a support core designed for removal from inside the flexible tube. During installation, the supporting core collapses on demand to allow the tube to shrink in order to contact with a wire or cable that needs protecting. Therefore, compared to the traditional heat shrink and pre-molded technology, the cold shrink splices have several advantages such as ease and repeatability of installation, i.e., no heat, no tools required, and broader cable size accommodation.

Conventional cold shrink products in the market are made of silicon rubber, such as high temperature vulcanization (HTV) and liquid silicone rubber (LSR), which have excellent tensile recovery performance due to their low intermolecular interaction and less molecular entanglement. However, there are also several drawbacks for such silicon-rubber-based cold shrink products such as low tear resistance and low alternating current breakdown strength. The cold shrink products may be damaged during demolding and expanding process if the tear resistance is not sufficient. In addition, end-users have reported tear failure during installation or usage.

It is believed that since the part remains stretched during its service life, a low tear strength results in rapid crack propagation if initiated by a sharp object during installation or pinches off from the surrounding environment during its service life. Also, the high alternating current breakdown strength is a key requirement in the cable accessories application for connection longevity.

Improved cold shrink materials with excellent tensile recovery performance, as well as low tear resistance and low alternating current breakdown strength, have been described (75). These compositions contain a silicone rubber, an ethylene/α-olefin interpolymer, a filler, and an α,β-unsaturated carboxylic acid.

Useful silicone rubbers, or silicone elastomers, are poly(organo siloxane)s, and preferably peroxidically crosslinking poly(organo siloxane)s. Representative silicone rubbers or poly(organo siloxane)s are collected in Table 4.6. Some of the monomers are shown in Figure 4.13.

Table 4.6 Silicone rubbers for cold shrink splices (75).

Compound
Dimethylvinylsiloxy-terminated poly(dimethyl siloxane)
Trimethylsiloxy-terminated poly(dimethyl siloxane)
Trimethylsiloxy-terminated copolymer of methylvinyl siloxane and dimethyl siloxane
Dimethylvinylsiloxy-terminated copolymer of methylvinyl siloxane and dimethyl siloxane
Dimethylhydroxysiloxy-terminated poly(dimethyl siloxane)
Dimethylhydroxysiloxy-terminated copolymer of methylvinyl siloxane and dimethyl siloxane
Methylvinylhydroxysiloxy-terminated copolymer of methylvinyl siloxane and dimethyl siloxane
Dimethylhexenylsiloxy-terminated poly(dimethyl siloxane)
Trimethylsiloxy-terminated copolymer of methylhexenyl siloxane and dimethyl siloxane
Dimethylhexenylsiloxy-terminated copolymer of methylhexenyl siloxane and dimethyl siloxane
Dimethylvinylsiloxy-terminated copolymer of methylphenyl siloxane and dimethyl siloxane
Dimethylhexenylsiloxy-terminated copolymer of methylphenyl siloxane and dimethyl siloxane
Dimethylvinylsiloxy-terminated copolymer of methyl(3,3,3-trifluoropropyl)siloxane and dimethyl siloxane
Dimethylhexenylsiloxy-terminated copolymer of methyl(3,3,3-trifluoropropyl)siloxane and dimethyl siloxane

Most preferably, the silicone rubber is a vinyl-terminated silicone rubber such as Silastic GP-30 from Dow Corning and silicone rubber 110 series available from Dongjue Silicone Group, China (75).

The ethylene-based polymers used here contain at least one ethylene/α-olefin interpolymer, which may optionally contain a diene.

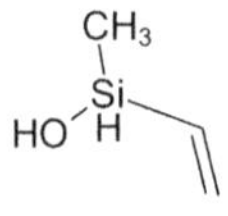

Methylvinyl siloxane

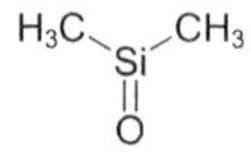

Dimethyl siloxane

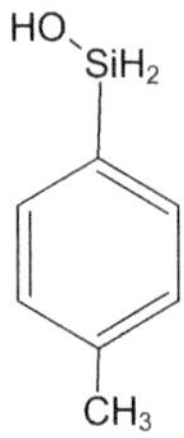

Methylphenyl siloxane

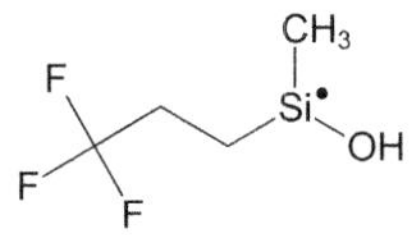

Methyl(3,3,3-trifluoropropyl)siloxane

Figure 4.13 Siloxane monomers.

As curing agent, any peroxide that will promote the crosslinking of the composition can be used. Exemplary peroxides include dicumyl peroxide, bis(*α*-*tert*-butyl peroxyisopropyl)benzene, isopropylcumyl-*tert*-butyl peroxide, *tert*-butylcumylperoxide, di-*tert*-butyl peroxide, 2,5-bis(*tert*-butylperoxy)-2,5-dimethylhexane, 2,5-bis(*tert*-butylperoxy)-2,5-dimethylhexane, 1,1-bis(*tert*-butylperoxy)-3,3,5-trimethylcyclohexane, isopropylcumyl cumylperoxide, di(isopropylcumyl) peroxide, or mixtures from these peroxides.

Some peroxides are shown in Figure 4.14.

Scorch inhibitors include 2,2,6,6-tetramethylpiperidinoxyl and 4-hydroxy-2,2,6,6-tetramethylpiperidinoxyl, c.f. Figure 4.15.

Suitable antioxidants are collected in Table 4.7. Some antioxidants are shown in Figure 4.16.

4.14.3 *Lubricious Cable Jackets for Medical Uses*

Silicone rubber is widely used throughout the medical industry as a cable jacket and as an overmold material for forming flex relief structures thereon (76). Reusable surgical device cables require excellent bonding and resistance to high temperatures and chemical steriliza-

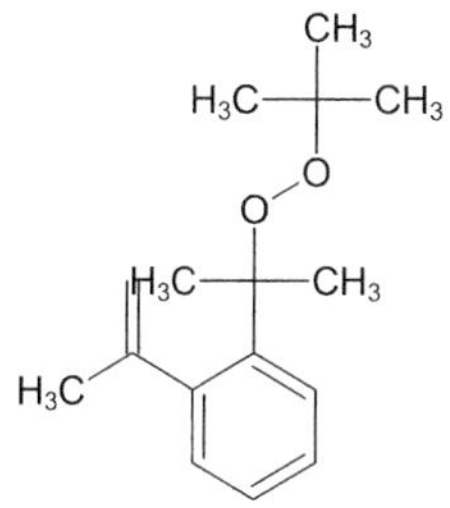

Isopropylcumyl-*tert*-butyl peroxide

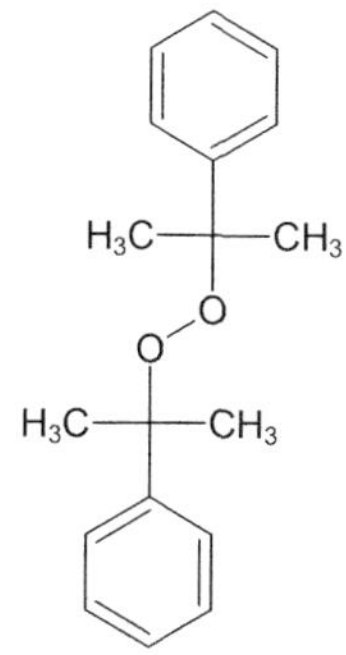

Dicumyl peroxide

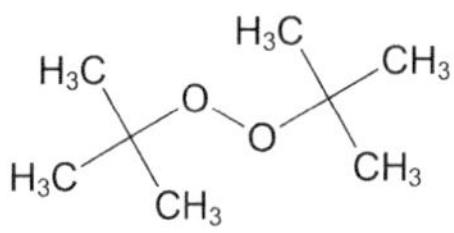

Di-*tert*-butyl peroxide

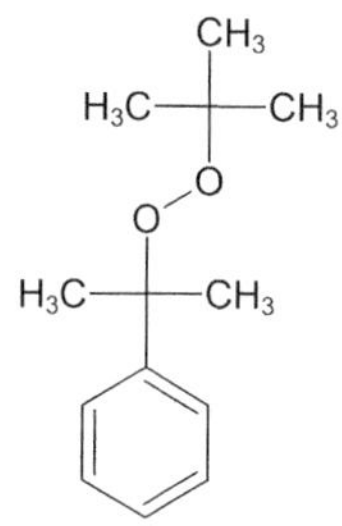

tert-Butylcumylperoxide

Figure 4.14 Peroxides.

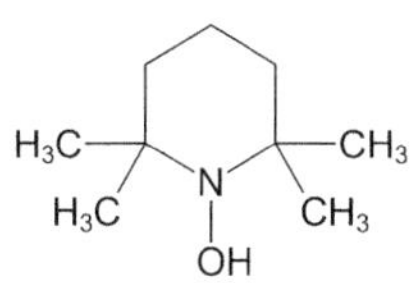

2,2,6,6-Tetramethylpiperidinoxyl

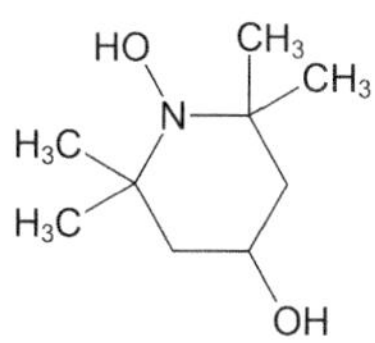

4-Hydroxy-2,2,6,6-tetramethylpiperidinoxyl

Figure 4.15 Scorch inhibitors.

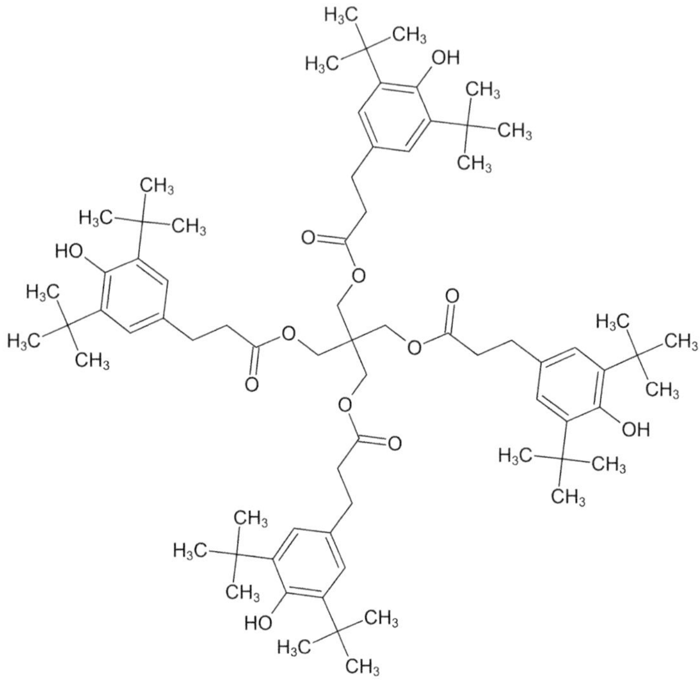

Figure 4.16 Tetrakis[methylene(3,5-di-*tert*-butyl-4-hydroxyhydro-cinnamate)]methane.

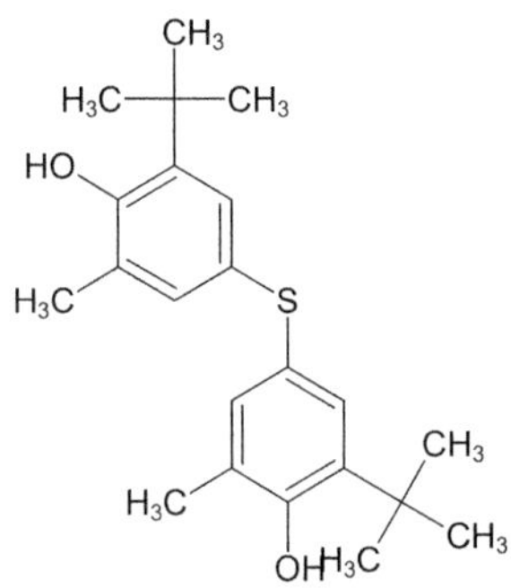

Figure 4.16 (cont.) 4,4′-Thiobis(2-methyl-6-*tert*-butylphenol).

Table 4.7 Antioxidants for cold shrink splice compositions (75).

Compound
Tetrakis[methylene(3,5-di-*tert*-butyl-4-hydroxyhydro-cinnamate)]methane
Bis[(β-(3,5-di-*tert*-butyl-4-hydroxybenzyl)methylcarboxy-ethyl)]-sulfide
4,4′-Thiobis(2-methyl-6-*tert*-butylphenol)
4,4′-Thiobis(2-*tert*-butyl-5-methylphenol)
2,2′-Thiobis(4-methyl-6-*tert*-butylphenol) and thiodiethylene
Bis(3,5-di-*tert*-butyl-4-hydroxy)-hydrocinnamate
Tris(2,4-di-*tert*-butylphenyl)phosphite
Di-*tert*-butylphenyl-phosphonite
Dilaurylthiodipropionate
Dimyristylthiodipropionate
Distearylthiodipropionate 2,2,4-Trimethyl-1,2-dihydroquinoline
n,*N*′-Bis(1,4-dimethylpentyl-*p*-phenylenediamine)
Alkylated diphenylamines
4,4′-Bis(α,α-dimethylbenzyl)diphenylamine
Diphenyl-*N*-phenylenediamine

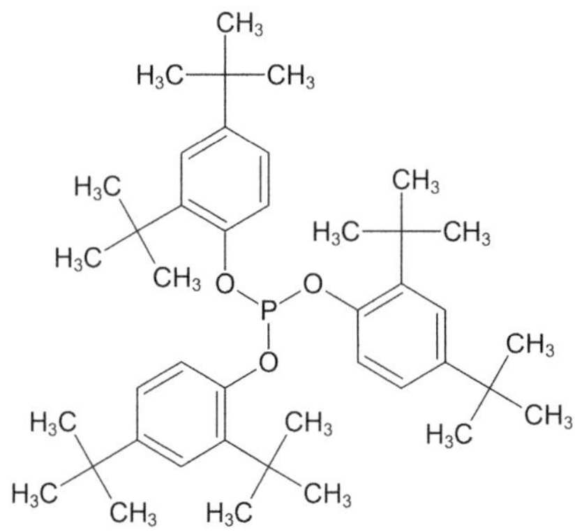

Figure 4.16 (cont.) Tris(2,4-di-*tert*-butylphenyl)phosphite.

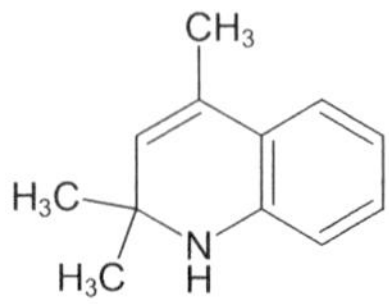

Figure 4.16 (cont.) 2,2,4-Trimethyl-1,2-dihydroquinoline.

tion techniques and equipment commonly used in hospitals, such as autoclaves.

The resistance to extreme heat and high humidity encountered during sterilization may also be relatable to the environmental requirements for other applications in the automotive, aerospace, and marine industries. Silicone rubber offers excellent chemical and thermal resistance properties as well as biocompatibility that greatly surpass the capabilities of other materials commonly used in cable jackets, such as poly(urethane) (PU) or poly(vinyl chloride).

However, silicone rubber also is naturally *tacky* to the touch and offers only a low level of abrasion resistance. End users in hospital environments are critical of the surface feel for many reasons (76):

1. Since cable assemblies are most often hand-held during surgical procedures, tackiness is a highly undesirable property.
2. Sterilization and reprocessing of cable assemblies require that cable jackets be routinely cleaned. A tacky cable jacket will attract dust and dirt, resulting in surgical devices that are difficult to clean and sterilize.
3. The cut resistance of a cable jacket is also extremely important when used in an operating room due to the presence of sharp surgical tools, i.e., scalpels.

A silicone cable jacket has been developed that exhibits lubricious properties (76). The silicone elastomer is a high consistency rubber or an LSR. The cable jacket may further contain a layer of an LSR that at least partially encapsulates the silicone elastomer and is bonded on it.

The properties were tested with some compositions. The silicone elastomer used for the samples consisted essentially of a commercial high consistency rubber resin. A poly(methylsilsequioxane) resin with an average particle size of between about 4 μm to 6 μm was incorporated as additive in the samples. The preparation and testing has been described in great detail (76). The following properties of the samples were tested: Abrasion, coefficient of friction, tear resistance, autoclave conditioning, and de-bonding.

It could be demonstrated that the loading of silicone elastomers with a poly(silsesquioxane) resin as an additive improves the properties of the compositions (76).

4.15 Electrophotography

4.15.1 Electrophotographic Fixing Device

A fixing device for an electrophotographic image-forming apparatus has been developed (77). This can be used for a copying machine, an electrophotographic printer, or an electrophotographic member.

A liquid silicone rubber mixture is used for the production of the electrophotographic member. The process of electrophotographic member formation includes the following steps (77):

1. Forming a layer of a liquid silicone rubber mixture on the substrate,
2. First crosslinking the liquid silicone rubber mixture in the layer of the liquid silicone rubber mixture to form a layer containing a cured silicone rubber and an agglomerate of the hollow particles dispersed in the cured silicone rubber,
3. Next crosslinking the cured silicone rubber, and
4. Causing the hollow particles in the layer containing the cured silicone rubber to undergo foam breakage to couple a plurality of holes derived from the hollow particles to each other to form the communication hole, and peeling the shells of the hollow particles from a wall surface of the cured silicone rubber to form the elastic layer.

The liquid silicone rubber mixture contains the following ingredients (**?**):

1. An organopolysiloxane having two or more unsaturated aliphatic groups in a molecule thereof,

2. An organohydrogenpolysiloxane having two or more hydrogen atoms bonded to a silicon atom in a molecule thereof,
3. A hydrosilylation catalyst,
4. Hollow particles each having a layer containing a thermoplastic resin as a shell,
5. A compound in which has a hydroxyl group and an ether bond, and having a sum of a number of the hydroxyl group and a number of the ether bond of two or more, and
6. A poly(carboxylic acid).

Specific examples of the ingredients are shown in Table 4.8.

Table 4.8 Specific examples of ingredients (77).

Component type	Material
Hollow particles	Acrylonitrile resin microballoons Matsumoto Microsphere FN-100MD
	Acrylonitrile resin microballoons Matsumoto Microsphere F-80DE
	Acrylonitrile resin microballoons Expancel 920DE80d30
Agglomerating agent	Tetraethylene glycol dimethyl ether Hisolve MTEM, boiling point: 275°C
	Triethylene glycol dimethyl ether Hisolve MTM, boiling point: 216°C
	Diethylene glycol dimethyl ether Hisolve MDM, boiling point: 162°C
	Triethylene glycol butyl methyl ether Hisolve BTM, boiling point: 261°C
	Tetraethylene glycol, boiling point: 314°C
	Triethylene glycol monomethyl ether Himol TM, boiling point: 249°C
Carboxylic acid	Citric acid
	DL-Malic acid
	DL-Tartaric acid
	Succinic acid
	o-Phthalic acid

4.15.2 *Electrophotographic Copy Machine*

In a final stage of printing or copying in image-forming apparatuses, such as an electrophotographic copy machine, a fax machine, or a laser beam printer, a heat fixing method is generally employed in which an image is fixed by bringing a pressure roller into contact with a fixing sleeve, fixing tube, or fixing roller having a heating source therein under pressure, and causing a transfer-receiving material having a transferred toner image thereon to pass between, thus melting a toner by heating (78).

4.15.2.1 *Sponge Fixing Roller*

A rubber-covered roller including a core serving as a roller base and a rubber layer for imparting elasticity, the rubber layer being disposed on the core, is widely used as a pressure roller. A known example of the rubber layer for imparting elasticity is composed of a rubber sponge, in particular, a silicone rubber sponge, containing pores (air bubbles) dispersed therein.

A silicone rubber sponge containing air bubbles dispersed therein can be produced, for example, by a method which disperses a foaming agent in a silicone rubber, and then curing the silicone rubber while foaming the foaming agent. However, according to this method, the resultant rubber surface may become uneven because, for example, foamed cells are exposed through a wall of the silicone rubber or an irregular foaming stress is generated in the silicone rubber and the stress is released. When depressions are formed on the rubber surface, the depressions remain on the surface even in the case where a releasing layer is formed on the rubber surface, which may result in a problem of stains forming on the pressure roller due to a molten toner, etc.

To address the above problem, a method for producing a silicone rubber sponge containing air bubbles dispersed therein has also been studied in which unexpanded resin microballoons are mixed in a liquid silicone rubber and the rubber is then cured by heating while expanding the resin microballoons. However, this method tends to cause a problem of unevenness of air bubbles. Consequently, a method has also been proposed in which (expanded) resin microballoons that have been expanded in advance are mixed in a liquid silicone rubber and the silicone rubber is then cured at a temperature equal to or lower than the melting point of the resin. For example, PTL 1 discloses a silicone rubber sponge containing air bubbles (pores) formed by expanded resin microballoons dispersed therein, and a pressure roller including an elastic layer composed of this silicone rubber sponge.

This elastic layer is formed by applying, onto a core, a mixture prepared by dispersing expanded resin microballoons in a liquid silicone rubber, and curing the liquid silicone rubber by conducting heating at a temperature lower than a softening point of the resin microballoons, and then breaking the resin microballoons after curing. This is because when shells of the resin microballoons remain in the resulting silicone rubber sponge, the shells may be broken as a result of thermal history during use of the silicone rubber sponge as a pressure roller and thus the fixing performance may vary. Thus, this problem is solved by breaking the shells in advance. This silicone rubber sponge is widely used as a pressure roller in an apparatus for forming a monochrome image.

A silicone rubber sponge has been developed that forms an elastic layer of a pressure roller of an image-forming apparatus (78). The silicone rubber sponge is capable of being used in a high-speed image-forming apparatus or a color image-forming apparatus and has a high mechanical strength.

The silicone rubber sponge is characterized in that a silicone rubber compound in which resin microballoons are dispersed is produced by the following steps (78):

1. A step of blending resin microballoons with a low-molecular-weight silicone rubber and mixing the resulting mixture to prepare a compound.
2. A step of mixing a silicone rubber having a molecular weight higher than that of the low-molecular-weight silicone rubber with the compound prepared in Step 1.

The silicone rubber sponge may further contain a reinforcing filler, an extending filler, a heat resistance additive, a colorant, a dispersion aid, an electrically conductive agent, a charge-controlling agent and a filler for adjusting thermal conductivity, as long as the object is not impaired. For example, the silicone rubber sponge may contain silica or metal silicon in order to improve thermal conductivity.

Figure 4.17 includes cross-sectional views illustrating the production process. Reference number 1 in Figure 4.17(a) denotes a metal cylinder (cylindrical metal mold). First, a PFA is applied onto the inner surface of the cylindrical metal mold and then cured to form a PFA layer 2. The PFA layer 2 functions as a surface releasing layer after the formation of a rubber-covered roller.

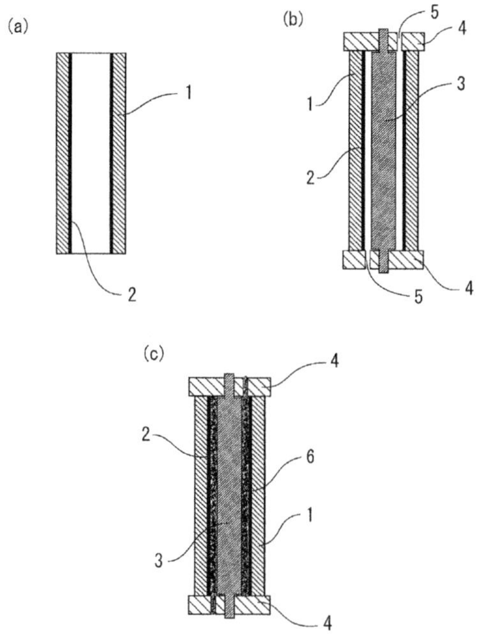

Figure 4.17 Cross-sectional views of the production process (78).

2	PFA layer	1	Cylindrical metal mold
3	Core	4	Cover
5	Injection inlet	6	Rubber elastic layer

After the perfluoroalkyl vinyl ether copolymer resin (PFA) layer is formed on the inner surface of the cylindrical metal mold, a core is arranged at the center of the inner space of the cylindrical metal mold, as illustrated in Figure 4.17 (b). The core is a column composed of a metal. Besides such a column composed of a metal, a core composed of a material that is generally used as a core of a pressure roller, for example, a resin having high heat resistance, chemical resistance, mechanical strength, etc., such as a polyimide resin, can also be used (78).

Openings at both ends of the cylindrical metal mold are closed by the covers 4. Each of the covers has a hole for holding an end of the core 3, the hole being formed at the center of the cover. The core is held by the holes at the center of the inner space of the cylindrical metal mold, and a gap having a certain distance is formed between the outer periphery of the core and the inner periphery of the cylindrical metal mold (PFA layer).

After the openings at both ends of the cylindrical metal mold are closed by the covers 4, a silicone rubber compound is injected into the gap. An inlet for injecting the silicone rubber compound is provided in each of the covers. Here, the silicone rubber compound is prepared by mixing expanded resin microballoons with a low-molecular-weight silicone rubber and uniformly dispersing the expanded resin microballoons to prepare a compound A, and further mixing a high-molecular-weight silicone rubber with the compound A.

Before the silicone rubber compound is injected, a primer for improving the adhesive strength between the PFA layer (surface releasing layer) and a silicone rubber (rubber elastic layer) may be applied onto the surface (inner surface) of the PFA layer, and dried. A primer for improving the adhesive strength between the silicone rubber (rubber elastic layer) and the core may be applied onto the surface of the core, and dried.

Figure 4.17 (c) illustrates a state where the silicone rubber compound is injected from the inlets into the gap. After the silicone rubber compound is injected, the whole cylindrical metal mold is heated to cure the silicone rubber compound (primary vulcanization), thus forming a rubber elastic layer containing the resin microballoons dispersed therein. The heating is conducted at a temperature lower than the softening point of the outer shells of the resin microballoons. If the heating temperature is equal to or higher than the softening point, thermal deformation and breaking of the balloons may occur, and a uniform sponge foam may not be formed. In the case where a commercially available product, such as Matsumoto Microsphere F series, is used as the resin microballoons, this heating is preferably performed at about 150°C for about one hour.

After the formation of the rubber elastic layer, the covers and the cylindrical metal mold are removed (demolded). Thus, a rubber-covered roller is produced (78).

4.15.2.2 Fixing Roller

Image-forming apparatuses, such as electrophotographic copying machines and laser beam printers, typically adopt a heat fixing method that forms an image of toner on recording paper with an image-forming process and thermally fixes the image (79). The toner is made of, for example, a heat-melting resin. The image-forming process includes electrophotography, electrostatic recording, and magnetic recording.

Conventional heat-fixing methods involve a roller for fixing or pressuring. The roller often includes a metal core shaft made of, for example, aluminum, a rubber layer on the shaft, made of silicone rubber, and a fluorine resin layer on the rubber layer, made of fluoroplastics with good toner releasability, such as a tetrafluoroethylene-perfluoroalkyl vinyl ether copolymer or poly(tetrafluoroethylene).

In recent years, a roller has been increasingly used which has a rubber layer formed of foam rubber (sponge rubber), in particular, open-cell foam rubber, from viewpoints of sufficient nip width, energy saving and a reduction of volatile organic compounds (VOC). The open-cell foam rubber has higher air permeability and prevents the outside diameter and the surface hardness of the roller from varying due to a rise in temperature of the roller in operation. The variation of the surface hardness and the outside diameter vary a fixed image or a paper feed speed.

A method has been developed that is capable of forming foam rubber with good uniformity and a high open-cell ratio, thereby capable of manufacturing a heat-fixing rubber roller having good fixing performance at a low cost and stably, and a rubber roller was manufactured by the method.

The method of manufacturing a heat-fixing rubber roller includes (79):

1. Forming a rubber layer of a silicone rubber composition on the outer periphery of a metal core shaft, the composition containing water-soluble sugar powder and triethylene glycol,
2. Vulcanizing the rubber layer, and
3. Eluting the sugar powder and the triethylene glycol from the vulcanized rubber layer to form a foam rubber layer.

Examples of the sugar powder are glucose, lactose, sucrose, trehalose, fructose, sugar, granulated sugar, and powdered sugar. Among these, granulated sugar and powdered sugar have high solubility in water and thus are better suited as the sugar powder. To obtain a sugar powder whose particle size is 10 μm to 100 μm a sieve sorting method can be used. The content of the sugar powder in the silicone rubber composition is most preferably 100 to 150 parts by mass per 100 parts by mass of the base silicone rubber.

The triethylene glycol contained with the sugar powder in the silicone rubber composition is a component that facilitates the elution of the sugar powder from the vulcanized rubber layer, enabling the formation of a foam rubber layer having a high open-cell ratio. Specifically, triethylene glycol has low compatibility with silicone rubber and high compatibility with sugar powder. The triethylene glycol mixed with the sugar powder in the silicone rubber forms a layer around the sugar powder with which it has high compatibility to prevent a skin layer of the silicone rubber from being formed between the sugar powder and the sugar powder. This facilitates eluting the sugar powder from the vulcanized silicone rubber layer, and the triethylene glycol itself is easily eluted from the silicone rubber layer, enabling formation of the foam rubber layer having a high open-cell ratio (79).

An assembly for the preparation is shown in Figure 4.18 and the method is detailed below.

The appropriate temperature of the primary vulcanization depends on the kind of the silicon rubber used. When the base silicone rubber is, for example, addition reaction LSR, the temperature of the primary vulcanization is most preferably within a 100°C to 130°C range.

After the primary vulcanization of the silicone rubber composition, the cylindrical mold is opened and the metal core shaft having the silicon rubber layer thereon is taken out. Specifically, the upper plug and the lower plug are detached and the metal core shaft together with the silicon rubber layer is extracted from the cylinder. Next, the extracted metal core shaft is immersed in water. Upon immersion, the silicone rubber layer is permeated with the water, and the sugar powder and the triethylene glycol dispersed in the rubber layer are dissolved in the permeated water to be discharged out of the rubber layer together with the water. To promote the

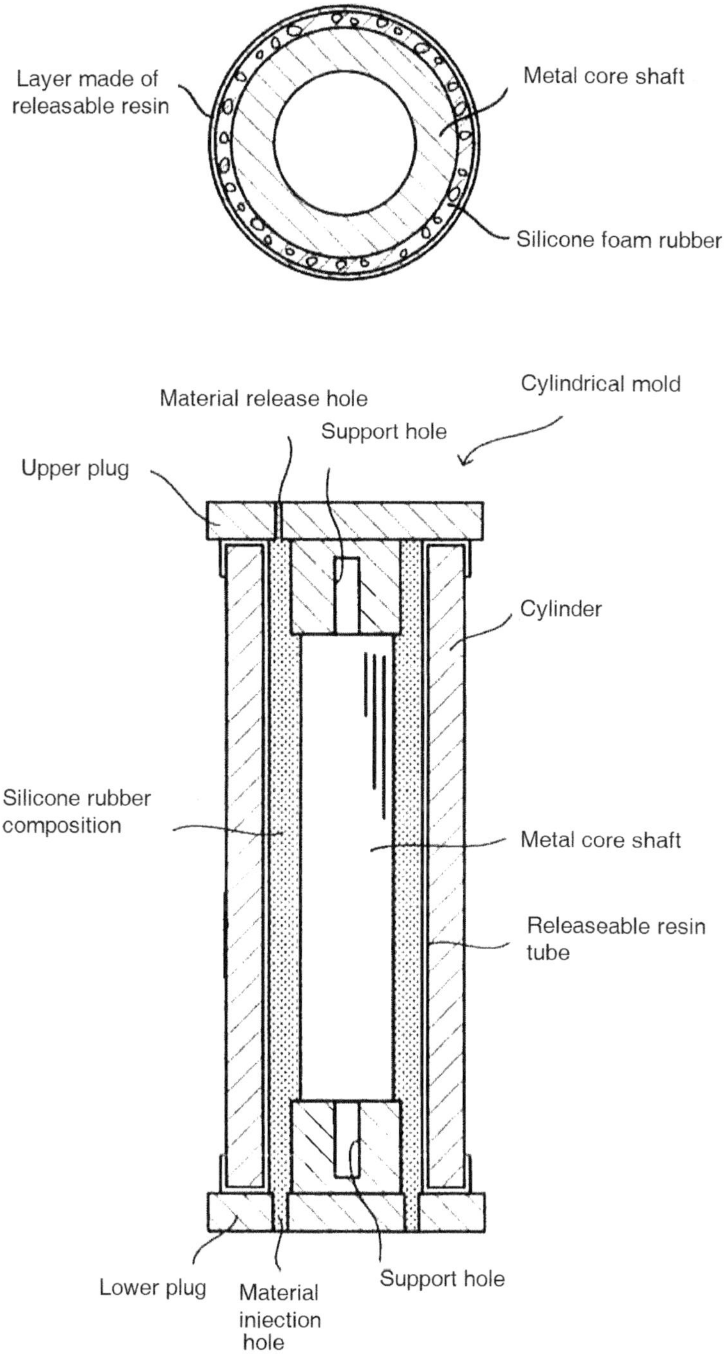

Figure 4.18 Assembly for preparation (79).

elution of the sugar powder and the triethylene glycol from the rubber layer, the water is preferably hot water at about 60 to 90°C, for instance. The immersion time is normally about one to eight hours, although this depends on the thickness of the rubber layer or the temperature of the immersion water. Using an immersion tank including a mechanism which continuously or intermittently supplies and discharges the water can better promote the elution of the sugar powder and the triethylene glycol.

Thereafter, the metal core shaft 121 including the silicone rubber layer, from which the sugar powder and the triethylene glycol have been eluted, is taken out from the water and the silicone rubber layer is secondarily vulcanized. An appropriate temperature of the secondary vulcanization depends on the kind of the silicone rubber used, similarly to the temperature of the primary vulcanization. Specifically, when the base silicone rubber is, for example, the LTV silicone rubber, the temperature is preferably within a 180 to 230°C. range, and more preferably within a 200 to 220°C. range. At least part of the triethylene glycol remaining in the rubber layer not discharged in the previous step is discharged in the process of the secondary vulcanization (79).

The above-described method can form the silicone foam rubber layer having a high open-cell ratio of 90% or more because the triethylene glycol compounded together with the sugar powder in the silicone rubber composition promotes the discharge of the sugar powder dispersed in the silicone rubber layer to the outside. The heat-fixing rubber roller including the silicone foam rubber layer with such a high open-cell ratio can stably form a good fixed image because of there being little change in its outside diameter and surface hardness due to a temperature increase during the operation, in addition to its low heat conductivity, which can contribute to a reduction of power consumption (79).

4.15.2.3 Electroconductive Member

In an image-forming process of an electrophotographic apparatus, a photoconductor drum serving as an electrostatic latent image-bearing member is uniformly charged with a charging member (80). An electrostatic latent image is formed on the photoconductor drum by the dissipation of charge due to exposure to a laser or the like. An appropriately charged developer in a developer container is uniformly applied to the developer roller using a developing roller and a regulating member configured to regulate the amount of the developer. The electrostatic latent image on the photoconductor drum is developed with a toner at the contact portion between the photoconductor drum and the developing roller. The toner on the photoconductor drum is transferred to a recording sheet with a transfer roller and fixed with a fixing device, thereby producing an electrophotographic image.

Figure 4.19 is a cross-sectional view of a process cartridge including an electroconductive member. The electroconductive member

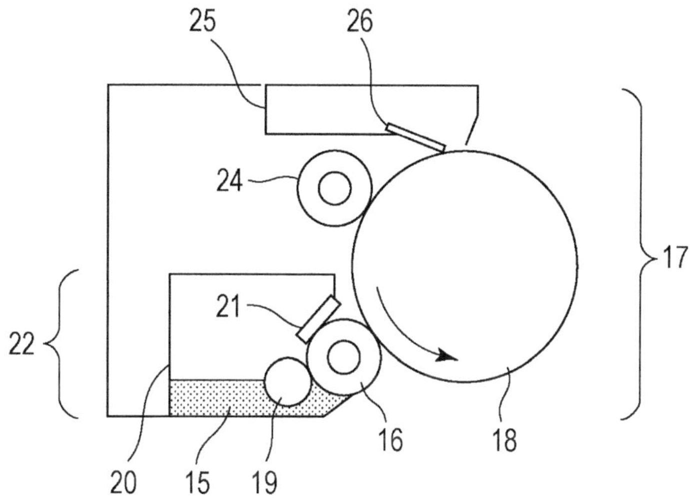

Figure 4.19 Process cartridge (80).

serves as a developing roller. Integrated with the process cartridge 17 are a developing roller 16, a developing blade 21, an electrophotographic photoconductor 18, a cleaning blade 26, a waste toner container 25, and a charging roller 24. The process cartridge is detachably attachable to the main body of the electrophotographic apparatus. A developing device 22 includes a toner container 20. The toner container 20 is charged with toner 15. The toner in the toner container 20 is supplied to a surface of the developing roller 16 with a toner supplying roller 19. A toner layer having a predetermined thickness is formed on the surface of the developing roller 16 with the developing blade 21.

The electrophotographic photoconductor 18 rotates in the direction indicated by an arrow and is uniformly charged with the charging roller 24 configured to subject the electrophotographic photoconductor 18 to charging treatment. An electrostatic latent image is formed on a surface of the electrophotographic photoconductor 18 with laser light 23 emitted from an exposure device configured to write an electrostatic latent image on the electrophotographic photoconductor 18. The electrostatic latent image is developed and visualized as a toner image by supplying the electrostatic latent image with toner using the developing device 22 arranged in contact with the electrophotographic photoconductor 18.

The toner left on the photoconductor drum is removed by a cleaning blade to complete the printing process.

In such an image-forming process, an electroconductive member for electrophotography, which is used as a charging member or developing member, typically includes an electroconductive elastic layer on an electroconductive base and a resin layer outside the elastic layer.

The electroconductive member is required to have the following characteristics (80): Appropriate conductivity, excellent endurance, and others. As an elastic layer of such an electroconductive member, a silicone rubber elastic layer in which deformation that is not easily recovered is less likely to occur even under long-term compression, namely, having excellent compression set properties, can be used. The silicone rubber elastic layer is produced by curing a silicone rubber mixture in which an electroconductive filler, such as carbon black, is added to an addition-curable liquid silicone rubber.

An electroconductive member for electrophotography has been developed that includes an electroconductive base, an electroconductive elastic layer, and a resin layer, in which the elastic layer contains a cured product of an addition-curable silicone rubber mixture containing the following components (80):

1. An organopolysiloxane that contains an alkenyl group attached to a silicon atom in its molecule,
2. An organopolysiloxane that contains a hydrogen atom attached to a silicon atom in its molecule,
3. Carbon black, and
4. Spherical solid inorganic particles.

4.16 Secondary Battery Pack

Batteries can be broadly classified into primary and secondary batteries. Primary batteries, also referred to as disposable batteries, are intended to be used until depleted, after which they are simply replaced with one or more new batteries. Secondary batteries, more commonly referred to as rechargeable batteries, are capable of being repeatedly recharged and reused, therefore offering economic, environmental and ease-of-use benefits compared to a disposable battery.

Examples of the secondary batteries may include nickel-cadmium batteries, nickel-metal hybrid batteries, nickel-hydrogen batteries, and lithium secondary batteries.

Lithium-ion batteries have emerged as a key energy storage technology and are now the main technology for consumer electronics devices, and industrial, transportation, and power-storage applications.

Due to their high potential and their high energy and power densities, and also their good lifetime, secondary batteries are now the preferred battery technology, in particular in the automotive industry, as it is now possible to provide longer driving range and suitable acceleration for electrically propelled vehicles such as hybrid electric vehicles, battery electric vehicles, and plug-in hybrid electric vehicles.

In the automotive industry, different sizes and shapes of lithium-ion battery cells are being manufactured and are subsequently assembled into packs of different configurations. An automotive secondary battery pack typically consists of a large number of battery cells to meet the desired power and capacity needs.

This switch in drive train technology is not, however, without its technological hurdles, as the use of an electric motor translates to the need for inexpensive batteries with high energy densities, long operating lifetimes, and the capability of operating in a wide range of conditions. Although rechargeable battery cells offer a number of advantages over disposable batteries, this type of battery is not without its drawbacks. In general, most of the disadvantages associated with rechargeable batteries are due to the battery chemistries employed, as these chemistries tend to be less stable than those used in primary cells.

Secondary battery cells, such as lithium-ion cells, tend to be more prone to thermal management issues which occur when elevated temperatures trigger heat-generating exothermic reactions, raising the temperature further and potentially triggering more deleterious reactions. During such an event, a large amount of thermal energy is rapidly released, heating the entire cell up to a temperature of 850°C or more. Due to the increased temperature of the cell undergoing this temperature increase, the temperature of the adjacent cells within the battery pack will also increase. If the temperature of these adjacent cells is allowed to increase unimpeded, they may also enter into an unacceptable state with exceedingly high temperatures within the cell, leading to a cascading effect where the initiation of temperature increases within a single cell propagate throughout the entire battery pack. As a result, power from the battery pack is interrupted and the system employing the battery pack is more likely to incur extensive collateral damage due to the scale of damage and the associated release of thermal energy.

In a worst case scenario, the amount of generated heat is great enough to lead to the combustion of the battery as well as materials in proximity to the battery. Furthermore, due to the characteristics of the lithium ion batteries, the secondary battery pack operates within an ambient temperature range of −20°C to 60°C. However, even when operating within this temperature range, the secondary battery pack may begin to lose its capacity or ability to charge or discharge should the ambient temperature fall below 0°C.

Depending on the ambient temperature, the life cycle capacity or charge/discharge capability of the battery may be greatly reduced as the temperature stays below 0°C. Nonetheless, it may be unavoidable that the lithium ion battery be used where the ambient temperature falls outside the optimum ambient temperature range, which is between 20°C to 25°C. These factors not only greatly shorten the driving range of the vehicle, but also cause great damage to the battery. Deterioration in energy and power available at lower temperatures is attributed to reduction in capacity and increase in internal resistance.

Alluding to the above, in a battery or battery assembly with multiple cells, significant temperature variances can occur from one cell to the next, which is detrimental to the performance of the battery pack. To promote long life of the entire battery pack, the cells must be below a desired threshold temperature. To promote pack performance, the differential temperature between the cells in the secondary battery pack should be minimized. However, depending on the thermal path to ambient temperature, different cells will reach different temperatures. Furthermore, for the same reasons, different cells reach different temperatures during the charging process.

Accordingly, if one cell is at an increased temperature with respect to the other cells, its charge or discharge efficiency will be different, and, therefore, it may charge or discharge faster than the other cells. This will lead to a decline in the performance of the entire pack.

A number of approaches have been employed to either reduce the risk of thermal issues, or reduce the risk of thermal propagation. These can be found in a patent (81), which discloses a battery pack thermal management system that comprises at least one enclosure failure port integrated into at least one wall of a battery pack enclosure, where the enclosure failure port(s) remains closed during normal operation of the battery pack, and opens during a battery pack thermal event, thereby providing a flow path for hot gas generated during the thermal event to be exhausted out of the battery pack enclosure in a controlled fashion. Another approach is to develop new cell chemistries and/or modify existing cell chemistries.

Yet another approach is to provide additional shielding at the cell level, thus inhibiting the flow of thermal energy from the cell undergoing thermal management issues propagating to adjacent cells. Still yet another approach is to use a spacer assembly to maintain the position of the battery undergoing the thermal event in its predetermined location within the battery pack, thereby helping to minimize the thermal effects on adjacent cells.

Thermally insulating a battery pack has also been described to reduce the risk of thermal excursions or their propagation. For example, a battery of lithium generators (82) has been described in which the generators are stacked one on top of another and this stack is held in position by being surrounded by PU foam. An embodiment is also disclosed in which cooling fins are inserted between two generators.

Also, a battery of lithium electrochemical generators has been described (83). This device includes: a casing, a plurality of lithium electrochemical generators housed in the casing, each generator including a container, a rigid, flame retardant foam with closed porosity formed of an electrically insulating material filling the space between the inner wall of the casing and the free surface of the side wall of the container of each electrochemical generator, the foam covering the free surface of the side wall of the container of each electrochemical generator over a length representing at least 25% of the height of the container.

According to one described embodiment, the foam consists of a material chosen from the group comprising PU, epoxy, polyethylene, melamine, polyester, formophenol, poly(styrene), silicone or a mixture thereof, PU and the mixture of PU and epoxy being preferred. The expansion of PU resin for foam-form is described using the following chemical routes to obtain the foam:

1. Via a chemical route, i.e., the reaction of water on isocyanate producing CO_2, which will cause the PU to foam,
2. Via a physical route, i.e., vaporization of a liquid with low boiling point under the action of heat produced by the exothermal reaction between isocyanate and the hydrogen-donor compound, or
3. Via injection of air.

However, rigid foams which are typically produced by reacting, for example, a polyisocyanate with an isocyanate reactive material, such as polyol, in the presence of a blowing agent do not exhibit the high compression set required when foams are used to minimize the adverse effect of any fire and explosion linked to a thermal event.

The main difference compared to traditional fuel-consuming vehicles is that there is no clutch or hydraulic torque converter in electric vehicles so the overall system configuration is inherently less elastic, as the motor and the transmission system are directly mechanically coupled. This configuration has little passive damping effect that can dampen disturbances and avoid oscillations, which are mostly noticeable during travel in the low speed range.

The dominating sound is the magnetic noise which generates a whining noise at high frequencies. A vehicle running only with an electric motor will also have less masking sound at low frequencies. This means that other noise requirements on, for example, component noise such as liquid or air cooling/heating for the electrical batteries must be changed accordingly. The noise during regeneration (battery charging) at coast down is also important. Therefore, due to the low damping in an electrical vehicle and lack of passive damping hardware as compared with a conventional vehicle, a damping control strategy is needed to minimize the drivetrain oscillations.

While a number of approaches have been adopted to try to lower the risk of thermal incursions as well as thermal energy propagation throughout the battery pack, it is critical that if a pack-level thermal event does occur, personal and property risks are minimized. As the number of cells in a battery increases, and as the size of the cells increases, so does the necessity and benefit of providing suitable thermal management.

Furthermore, there is still a need to better insulate battery cells, in particular lithium-ion batteries, from the adverse effect of low temperature that are met when the weather reaches severe low temperature that could reach −20°C and even lower.

A novel silicone rubber syntactic foam and the silicone precursor of said silicone rubber syntactic foam has been developed, which could be used in secondary battery pack and that will minimize personal and property risks due to uncontrolled thermal events still awaiting (84).

It is thought that the silicone rubber syntactic foam and the silicone precursor of the silicone rubber syntactic foam will address the above-mentioned problems linked to uncontrolled thermal excursions, in particular for lithium batteries, and will present efficient low temperature insulation properties and provide a damping control strategy to minimize the drivetrain oscillations.

An addition curing type organopolysiloxane composition is used that contains a reactive diluent for reducing the viscosity of the composition, a hydrosilylation catalyst, hollow glass beads and hollow borosilicate glass microspheres (84).

4.17 Pressure and Temperature Sensor

A pressure/temperature sensitive SR filled with carbon black was prepared by a liquid mixing method (85).

A room-temperature vulcanized liquid silicone rubber from Sichuan Chenguang Engineering Design Institute, China, was used as a matrix. Carbon black without any chemical functional groups attached, was used as a conductive filler. For carbon black, the pour density was 128 $g\,dm^{-3}$, the iodine adsorption was 1080 $mg\,g^{-1}$, the cetyltrimethylammonium bromide surface area was 1100 $m^2\,g^{-1}$, and the dibutyl phthalate absorption was 3,8 $ml\,g^{-1}$. A volume of 3 *ml* of naphtha was used as a solvent to mix the fillers with the rubber. This was followed by mechanical stirring along with sonication for 15 *min* to achieve a uniform dispersion. The viscous mixture was modeled onto the cylinder, which was placed in a vacuum oven for 72 *h* to reduce the air porosity. After the solvent evaporated, the samples were manufactured into squares of (17×17 mm^2) at 258°C.

For a testing apparatus for the pressure/temperature experiment, two copper plates used as electrodes were placed on the worktable. The area of the electrode was a little larger than that of the sample to retain good electrical contact. Moreover, silver paint was used to ensure good contact of the sample surface with the electrodes to reduce the contact resistance. The pressure experiment was provided by a digital force gauge with a precision of 0.003 *MPa* within the pressure range from 0 to 1 *MPa*. The temperature experiment was provided by a temperature-programmable chamber with a precision of 0.58°C within the temperature range from room temperature (25°C) to 80°C. The values of the resistor were recorded with an interval of 5°C.

The transfer function of a pressure/temperature sensor based on carbon black/SR was derived by general effective media theory. The results showed that the transfer functions coincided well with the experimental data, and the negative pressure coefficient of the resistance/positive temperature coefficient of resistance was shown.

The working principles of these two kinds of sensors are different. The working principle of the pressure sensor based on carbon black/SR was related to the volume fraction of carbon black.

With increasing volume fraction of carbon black, the working principle of this kind of pressure sensor varied from a piezoresistive effect to a strain effect. In addition, the working principle of the temperature sensor based on carbon black/SR was that the resistivity changed with temperature; this was not related to the volume fraction of carbon black.

The pressure dependence of the resistance on the content of carbon black is shown in Figure 4.20 and the temperature dependence of the resistance on the content of carbon black is shown in Figure 4.21.

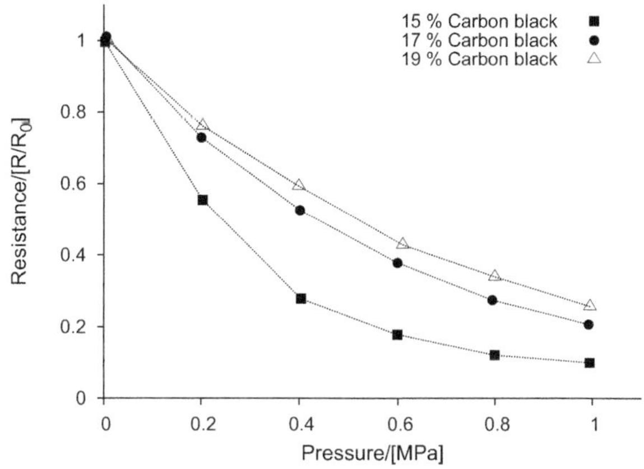

Figure 4.20 Pressure dependence of resistance (85).

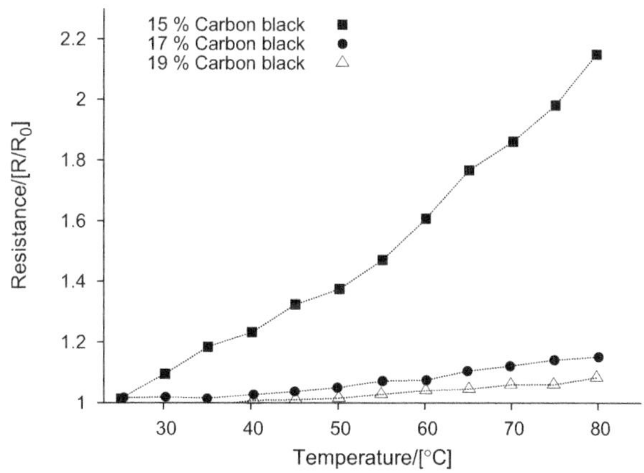

Figure 4.21 Temperature dependence of resistance (85).

4.18 Piezoresistive Device

A piezoresistive device with a carbon black particle network in a highly flexible silicone rubber for large deformation and wide range strain sensing has been investigated (86). The piezoresistive composite film was fabricated using a mixture of silicone rubber and carbon black filler particles.

Testing of the composition showed that the carbon black particle network in the silicone rubber strongly affected the resistance of the device during the process of drawing and its recovery. It was found that a 50% volume ratio of carbon black filler particles showed a lower relative resistance than a 33.3% volume ratio of carbon black filler particles, but with an advantage of good resistance recovery stability and a smaller perturbation error (smaller changed resistance) during the periodic back and forth linear motor test.

With having both a 50% volume ratio of carbon black filler particles and a 33.3% volume ratio of carbon black filler particles, up to 200% strain with resistances of 18 $k\Omega$ and 110 $k\Omega$ can be obtained, respectively (86).

Also, it was found that the relative resistance increased in an approximately linear relationship corresponding to the value of step-increased instantaneous length for these devices. An application test through hand drawing was used to demonstrate the piezoresistive performance of the device, which showed that the device was capable of measuring the instantaneous length with a large deformation (86).

In a related study, the relation between the pressure and the capacitance of the carbon black filled silicone rubber composite with a carbon black content of 5 to 9 *vol%* under multi-compression cycles was investigated (87).

Carbon black was used as the conductive phase of the composite. Room temperature vulcanized liquid silicone rubber was used as the polymer matrix. The capacitance of the composite increases with the increase of the pressure. There are deviations between the thicknesses/capacitivities of the composite under the same pressure in the different compression cycles, which results in the repeatability error of the piezocapacitivity of the composite.

The experimental results showed that the repeatability can be improved by increasing the number of the compression cycles. Also, it was demonstrated that the hysteresis error, the nonlinear error and the sensitivity decrease with the increase of the number of the compression cycles (87).

4.19 Proton Exchange Membrane Fuel Cells

Gaskets made of silicone rubber materials were compressed in the proton exchange membrane fuel cell to keep the gases and the oxidant in their own zones from leaking (88).

Silicone rubber is a promising candidate for gaskets in a proton exchange membrane fuel cell. In general, silicone is able to be crosslinked using peroxides, and the popular fillers are carbon black and calcium carbonate. It has already been widely used as gaskets in other applications due to its flexibility and resistance to weathering. It is stable in temperature from −40°C to 120°C. Its glass transition temperature at −54°C is relatively low and it can remain flexible, elastic, and retains its properties up to 150°C (89).

Although its surface chemistry may change over time when exposed to proton exchange membrane fuel cell environment, its bulk mechanical properties remain intact (88). However, like most polymers, silicone rubber also shows viscoelastic behavior, especially at high temperature.

The mechanical properties of the silicone rubbers were studied by exposing them to high temperature, humid air and acidic solution environment, which is similar to the real proton exchange membrane fuel cell operation condition (88).

Compression set and stress relaxation tests were performed. The results indicated that the temperature, the humid air, and the acidic solution have important effects on the mechanical properties of the silicone rubber materials. All three factors can induce the increase of the stress relaxation modulus and the compression permanent deformation. This can accelerate the aging of the mechanical properties and decrease the sealing property of the gaskets, which will influence the durability of the proton exchange membrane fuel cell (88).

4.19.1 Degradation Experiments

The degradation of silicone rubbers with different hardness in various aqueous solutions was investigated. It was found that silicone rubbers became more durable in aqueous solutions with the increase in hardness (90).

The degradation of seals during polymer electrolyte membrane fuel cell (PEMFC) operation was reported. It was found that the small particles released from silicone rubbers could migrate to the membrane electrode assembly and lower the electrochemical performance of the fuel cell (91).

The degradation of silicone rubbers exposed to PEMFC environment was studied (92–96). The results of these studies showed that silicone rubbers degraded more severely with increasing acid environment and temperature. The degradation of silicone rubbers was attributed to a chemical decomposition of the silicon-based backbone accompanied with a leaching of the fillers.

Also, the sealing force and thermal stress development of silicone rubbers under temperature cycling were studied (97,98). The thermal expansion or contraction was the major contributor to the compressive stress developed in the silicone rubbers seal.

The degradation of silicone rubbers exposed to different environments under alternating temperature cycling from −20°C to 90°C was investigated in detail (99).

Methylvinyl silicone rubbers with the hardness and thickness of 50 (Shore A) and 0.5 *mm*, respectively, were the investigated objects. All the tested samples were rectangular with dimensions of 10.0 *cm* in length and 4.0 *cm* in width.

Three solutions with different pH values and air were used as testing environment to study the degradation of the silicone rubbers. These were reagent grade acetic acid, sulfuric acid (98%), and hydrofluoric acid (48%). The first solution with a pH value of 3.35 simulates an actual PEMFCs working environment, with chemical composition of 12 *ppm* H_2SO_4 and 1.8 *ppm* HF with reagent grade. The second testing environment was acetic acid solution with a pH value of 5.0. The third liquid environment with a pH value of 7.0 was deionized water. The weight and hardness changes of the silicone rubber samples were monitored at every certain number of temperature cycles. The weight changes under temperature cycling are shown in Figure 4.22 and the Shore A hardness changes under temperature cycling are shown in Figure 4.22.

The SEM images of the silicone rubbers showed surface topographical changes of the silicone rubber after exposure to the acetic acid solution. Cracks already occurred on the surface of the sample after 50 cycles. Then the cracks or voids grew with the increase of

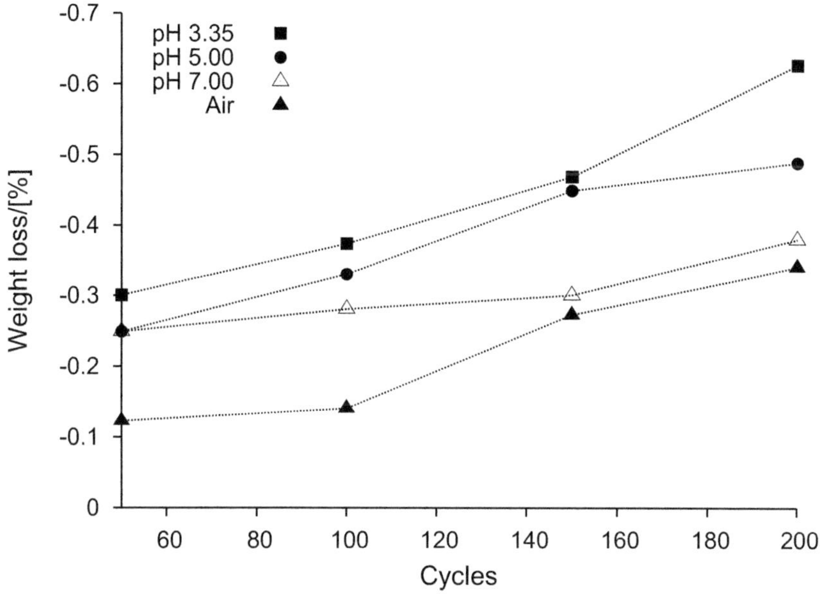

Figure 4.22 Weight changes under temperature cycling (99).

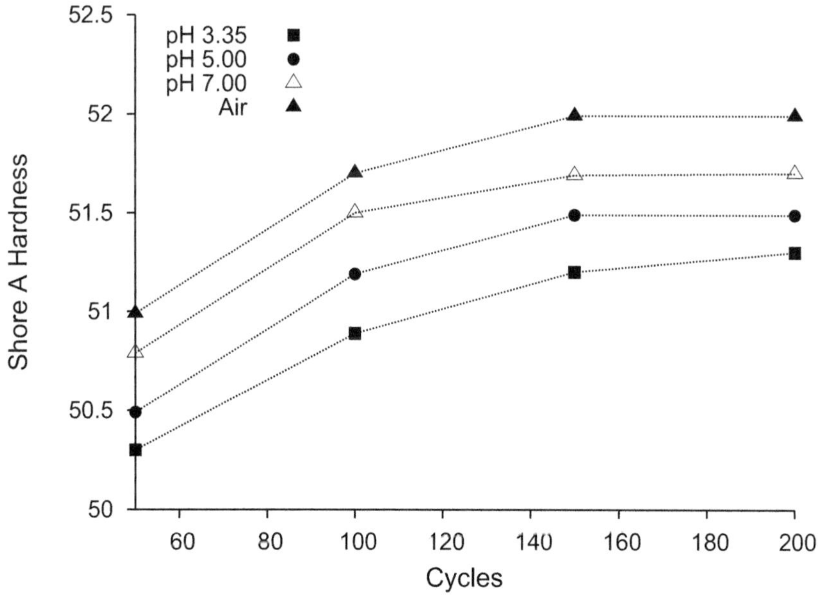

Figure 4.23 Shore A hardness under temperature cycling (99).

alternating temperature cycles. In the case of water, after 50 cycles the surface condition of the samples exposed to the testing environment did not change significantly. Cracks were gradually observed but the surface almost remained smooth. It was not until 100 cycles that voids or cracks formed on the surface of the samples exposed to the environments. In the case of air, the samples changed only a little and no voids or cracks were observed before 200 cycles.

So it can be concluded that the surface damage of the silicone rubber samples in the solutions is more serious than that in air. The increase in acidity and the temperature cycle numbers results in a significant damage of the silicone rubbers (99).

4.20 Light-Emitting Diodes

In the market of lighting technologies, LEDs have gradually replaced conventional light sources. Because of their high energy efficiency and long lifetime, they are increasingly used in consumer products, interior and exterior lighting applications in the home and mobility sector as well as in industrial applications. The material properties in the surrounding area of the light-emitting semiconductor chip are crucial to the performance of LED. Although the energy efficiency of LED is higher compared to conventional light sources, temperatures exceed about 150°C close to the semiconductor chip. Especially in combination with high amounts of blue ultraviolet (UV) radiation, the materials for encapsulation cannot meet the requirements and reduce the lifetime of an LED significantly.

Contrary to conventional materials, highly transparent LSR which can resist high temperatures as well as UV radiation, offers great freedom in design. This enables the combination of the encapsulation (primary optics) and the secondary optics in one component. The objective of an ongoing joint research project with various partners from the industry is the development of an innovative injection molding process for high precision optics in LED applications made of LSR, which is being analyzed at the Institute of Plastics Processing (IKV), Aachen, Germany. Therefore, the LED board is placed in the injection mold and overmolded with LSR. The goal is a highly integrated process with major emphasis on the reduction of components, mounting steps and costs. Furthermore, the combination of primary and secondary optics promises an improved effectiveness because losses in light power due to the transition of the primary and secondary optics are reduced (100,101).

In another study, phenyl hydrogen polysiloxanes of different structures were prepared (102). LED packaging silicone rubbers were made from these polysiloxanes. The hardness, elongation, tensile strength and anti-yellowing of the LED packaging silicone rubbers on lamps were characterized.

The results of the study showed that phenyl hydrogen polysiloxanes of resin structures had a lower molecular weight. The cured silicone rubbers exhibited higher hardness, tensile strength and crosslinking density, and a lower gas permeability. Phenyl hydrogen polysiloxanes with oil structures had higher molecular weight, and the cured silicone rubbers from these polysiloxanes exhibited a lower volume expansion coefficient, higher hot and cold impact cycles and less change in hardness during the aging process.

When a phenyl vinyl silicone resin was used in yellowing studies it turned brown, which indicated that the yellowing of the silicone rubbers was most probably due to the residual vinyl groups and was not related to the phenyl groups.

The results of the study further showed that the degree of change in hardness, yellowing, gas permeability, sulfuration, and volume expansion coefficient was reduced by using phenyl hydrogen polysiloxanes of resin oil structures as crosslinking agents, thus enabling the combined performance advantages of both resin structure and oil structure rubbers (102).

4.21 Recycling of Used Composite Electric Isolators

Recycling and reusing materials for the development of new products has been an important subject for several years (103). Finding new processing methods and ways to improve the properties of the final materials is a trending subject.

The morphological and structural characteristics of recovered silicone rubber from the envelope of used composite electrical insulators obtained using two different technological approaches has been investigated.

The samples were obtained by injection and extrusion processes. Then they were subjected to a vulcanization step (103). The internal structure was analyzed with SEM in order to assess the character of the internal structure of the resulting materials.

The studies showed that the character of the samples obtained using the injection process is discontinuous, large cavities forming between the recovered rubber and the binder. However, the samples obtained by extrusion method have proved a uniform microstructural aspect, the internalizing of the recovered rubber granolas into the matrix being much higher.

Thus, it was concluded that the extrusion method is a better processing alternative for recovered silicone rubber from the envelope of used composite electrical insulators in comparison to the injection method concerning the microstructural characteristics of the resulting materials (103).

4.22 Triboelectric Nanogenerator for Wearable Electronics

The rapid advancement of intelligent wearable electronics imposes the emergent requirement for power sources that are deformable, compliant, and stretchable (104). Power sources with these characteristics are difficult and challenging to achieve. The use of liquid metals as electrodes may provide a viable strategy to produce such power sources.

A liquid-metal-based triboelectric nanogenerator has been used by employing Galinstan as the electrode and silicone rubber as the triboelectric and encapsulation layer (104).

Galinstan is a brand name and a common name for a liquid metal alloy, whose composition is part of a family of eutectic alloys mainly consisting of gallium, indium, and tin (105).

The small Young's modulus of the liquid metal ensures the electrode remains continuously conductive under deformations, stretching to a strain as large as ~300% (104). The surface oxide layer of Galinstan effectively prevents the liquid Galinstan electrode from further oxidization and permeation into silicone rubber, yielding outstanding device stability.

Operating in the single-electrode mode at 3 Hz, the liquid-metal-based triboelectric nanogenerator with an area of $6 \times 3\ cm^2$ produces an open-circuit voltage of 354.5 V, transferred short-circuit charge of 123.2 nC, short-circuit current of 15.6 μA, and average power density of 8.43 $mW\ m^{-2}$, which represent outstanding performance values for triboelectric nanogenerators.

Furthermore, the liquid-metal-based triboelectric nanogenerator maintains stable performance under various deformations such as stretching, folding, and twisting. liquid-metal-based triboelectric nanogenerators in different forms, such as bulk-shaped, bracelet-like, and textile-like, are all able to harvest mechanical energy from human walking, arm shaking, or hand patting to sustainably drive wearable electronic devices (104).

4.23 Large Specific Surface Area Electrodes

An approach has been presented to produce superhydrophobic, surface-conducting silicone rubber with a strongly developed surface modified with multi-walled carbon nanotubes partially embedded in the silicone elastic matrix (106). The modification was achieved by the physical deposition of a carbon nanotube powder on a semi-liquid silicone rubber surface prior to its crosslinking.

The resulting biomimetic material displayed superhydrophobic properties (static wetting angle greater than 160°, a sliding angle of around 10°), as well as elevated electric surface resistance (surface resistivity of approximate 18 $k\Omega$). A piezoresistive hysteretic response with nonmonotonic change of the surface resistance accompanying substantial linear stretching was also demonstrated in the developed specimens displaying negative resistance change in a broad range of extension ratios, making them applicable as highly compliant, large-specific-area electrodes (106).

4.24 Casing

Electronic devices, such as notebook computers, may include casings made of metal. In order to improve the texture of the casing, an outer surface of the casing may be sprayed with paint (107). However, the texture of the paint sprayed on the casing is often less than satisfactory, and the paint is easily peeled off from the casing.

A casing for electronic devices includes a main body, a silicone rubber layer, and a bonding layer. The main body is made of a magnesium alloy. The bonding layer is bonded between the main body and the silicone rubber layer. The bonding layer includes about 12% to about 30% of a silane coupling agent.

Figure 4.24 shows a schematic of an electronic device. The electronic device includes a casing and at least one electronic element received in the casing. The electronic device can be a notebook computer. The casing includes a main body, a bonding layer, and a silicone rubber layer. The bonding layer is bonded between the main body and the silicone rubber layer (107).

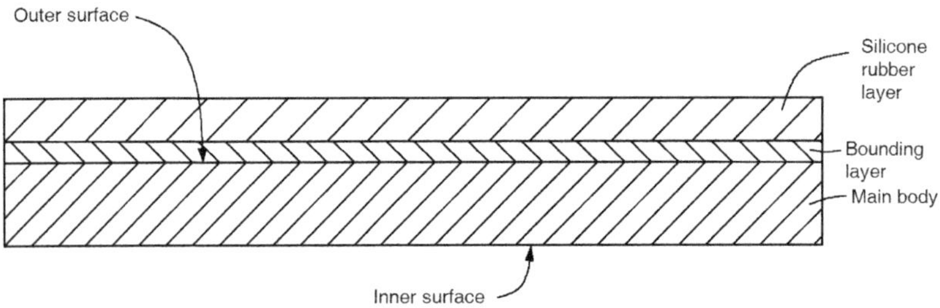

Figure 4.24 Electronic device (107).

The silicone rubber layer is electrically insulative, resistant to acid erosion, resistant to light damage, resistant to fungus, and chemically stable. The silicone rubber layer maintains high strength and high elasticity in high-temperature and low-temperature environments. A Shore hardness of the silicone rubber layer is about 10 to about 90, and a thickness of the silicone rubber layer is about 0.5 *mm* to about 2.0 *mm*.

A layer of liquid silicone rubber is sprayed on the bonding layer before the bonding layer is cured. The silicone rubber layer is formed on the bonding layer after the liquid silicone rubber is cured (107).

References

1. C. Freye and F. Jenau, DC conductivity measurements of liquid silicone rubber: Influence analysis and repeatability, in *2018 IEEE International Conference on Environment and Electrical Engineering and 2018 IEEE Industrial and Commercial Power Systems Europe (EEEIC / I CPS Europe)*, pp. 1–6, Budapest, Hungary, June 2018. IEEE International.
2. J. Kong, Y. Tong, J. Sun, Y. Wei, W. Thitsartarn, C.C.Y. Jayven, J.K. Muiruri, S.Y. Wong, and C. He, *Composites Science and Technology*, Vol. 159, p. 208, 2018.
3. M. Shigeta, T. Nagamune, and A. Katada, Silicone rubber composition and vulcanized product, US Patent Application 20 180 044 498, assigned to Zeon Corporation, Chiyoda-ku, Tokyo (JP), February 15, 2018.
4. H. Zhou, Z. Xia, X. Wang, Z. Li, T. Li, and F. Guo, *Polymer Composites*, Vol. 36, p. 1371, 2015.
5. A. Bunde and S. Havlin, eds., *Fractals and Disordered Systems*, Springer Science & Business Media, 2nd edition, 2012.
6. W.J. Chen, X. Zeng, X. Lai, H. Li, W.Z. Fang, and F. Hou, *ACS Applied Materials & Interfaces*, Vol. 8, p. 21039, 2016.
7. Y. Zhang, X. Zeng, X. Lai, H. Li, Q. Zhou, and X. Huang, *ACS Omega*, Vol. 2, p. 5111, 2017.
8. I. Ramirez-Vazquez and R. Hernandez-Corona, Tracking and erosion requirements for high voltage silicone rubber insulators, in *2016 IEEE International Conference on Dielectrics (ICD)*, Vol. 2, pp. 740–743, July 2016.
9. IEC, Selection guide for polymeric materials for outdoor use under HV stress, IEC 62039, International Electrotechnical Commission, Geneva, Switzerland, 2007.
10. S. Rowland, J. Robertson, Y. Xiong, and R. Day, *IEEE Transactions on Dielectrics and Electrical Insulation*, Vol. 17, p. 375, 2010.
11. K.O. Papailiou and F. Schmuck, *Silicone Composite Insulators: Materials, Design, Applications*, Springer Science & Business Media, 2012.
12. R. Chakraborty, *Studies on Silicone Rubber Insulators Used for High Voltage Transmission*. PhD thesis, Indian Institute of Science, Bangalore, India, 2017.
13. I.J.S. Lopes, S.H. Jayaram, and E.A. Cherney, *IEEE Transactions on Dielectrics and Electrical Insulation*, Vol. 8, p. 262, April 2001.
14. L.H. Meyer, S.H. Jayaram, and E.A. Cherney, *IEEE Transactions on Dielectrics and Electrical Insulation*, Vol. 11, p. 424, June 2004.
15. M.A. Douar, A. Beroual, and X. Souche, *IEEE Transactions on Dielectrics and Electrical Insulation*, Vol. 22, p. 391, February 2015.
16. M.A. Douar, A. Beroual, and X. Souche, *IET Generation, Transmission Distribution*, Vol. 10, p. 986, 2016.

17. N.A. Arshad, S.G. McMeekin, and M. Farzaneh, *Energies*, Vol. 9, 2016.
18. Y. Li, H. Jin, S. Nie, C. Tong, and N. Gao, *AIP Advances*, Vol. 8, p. 035117, 2018.
19. J. Van Loon, M. Schneider, P. Meier, and X. Kornmann, Method of producing a layer of a vulcanized silicone rubber composition having an improved adhesion to the substrate surface, US Patent 9 764 353, assigned to ABB Research Ltd. (Zurich, CH), September 19, 2017.
20. B. Grau, L. Bemert, H. Zumaque, and O. Safarowski, Use of a silicone rubber composition for the manufacture of an insulator for high voltage direct current applications, US Patent Application 20 170 372 815, assigned to Momentive Performance Materials GmbH, December 28, 2017.
21. E. Jonsson, T. Auletta, and H. Hillborg, Field grading material, WO Patent 2 008 076 058, assigned to ABB Research Ltd., June 26, 2008.
22. M. Dhofir, M.F. Aditama, R.N. Hasanah, and H. Suyono, *International Journal on Advanced Science, Engineering, Information Technology (IJASEIT)*, Vol. 7, p. 1340, 2017.
23. J. Pimper and R.F. Räder, Manufacturing process of a hollow composite insulator and hollow composite insulator, EP Patent 1 091 365, assigned to Cellpack AG, April 11, 2001.
24. J. Lambrecht, G. Simson, and H.-J. Winter, Method for producing composite insulators by UV-crosslinking silicone rubber, US Patent 9 236 164, assigned to Wacker Chemie AG (Munich, DE), January 12, 2016.
25. I. Kitta, S. Manjang, W. Tjaronge, and R. Irmawaty, *ARPN Journal of Engineering and Applied Sciences*, Vol. 11, p. 4689, 2016.
26. A. Jaya, H. Berahim, and R. Tumiran, *Electrical and Electronic Engineering*, Vol. 2, p. 208, 2012.
27. ASTM Subcommittee D09.12, Standard test methods for AC loss characteristics and permittivity (dielectric constant) of solid electrical insulation, ASTM D150-18, ASTM International, West Conshohocken PA, 2018.
28. H. Sun, D. Meng, B.-Y. Zhang, F. Wu, J. Chen, and H. Hillborg, Performance of liquid silicone rubber exposed to acid fog under DC stress, in *2015 IEEE Conference on Electrical Insulation and Dielectric Phenomena (CEIDP)*, pp. 229–232, October 2015.
29. B.S. Reddy and S. Prasad, *IEEE Transactions on Dielectrics and Electrical Insulation*, Vol. 22, p. 1711, 2015.
30. D.S. Prasad and B.S. Reddy, *IEEE Transactions on Dielectrics and Electrical Insulation*, Vol. 23, p. 1546, June 2016.
31. M.T. Nazir, B.T. Phung, and M. Hoffman, *IEEE Transactions on Dielectrics and Electrical Insulation*, Vol. 23, p. 2804, October 2016.

32. Z. Yan, X. Liang, C. Wu, W. Bao, S. Li, and Y. Liu, Aging and recovery of superhydrophobic silicone rubber under electrical and non-electrical stresses, in *2015 IEEE Conference on Electrical Insulation and Dielectric Phenomena (CEIDP)*, pp. 189–192, Ann Arbor, MI, USA, Oct 2015. IEEE.
33. A.R. Verma and B.S. Reddy, *IEEE Transactions on Dielectrics and Electrical Insulation*, Vol. 25, p. 46, February 2018.
34. ASTM Subcommittee D09.12, Standard test methods for liquid-contaminant, inclined-plane tracking and erosion of insulating materials, ASTM D2303-13, ASTM International, West Conshohocken PA, 2013.
35. IEC, Electrical insulating materials used under severe ambient conditions – test methods for evaluating resistance to tracking and erosion, IEC 60587, International Electrotechnical Commission, Geneva, Switzerland, 2007.
36. Y. Xia, X. Song, Z. Jia, and X. Wang, Mechanism and quantitative characterization of color fading phenomenon of HTV composite insulators silicone rubber, in *2018 12th International Conference on the Properties and Applications of Dielectric Materials (ICPADM)*, pp. 748–751, May 2018.
37. D. Ghosh, S. Bhandari, T.K. Chaki, and D. Khastgir, *RSC Adv.*, Vol. 5, p. 57608, 2015.
38. C. Zhu, C. Deng, J.-Y. Cao, and Y.-Z. Wang, *Polymer Degradation and Stability*, Vol. 121, p. 42, 2015.
39. B. Liu, X. Gao, Y. Zhao, L. Dai, Z. Xie, and Z. Zhang, *Journal of Materials Science*, Vol. 52, p. 8603, July 2017.
40. R.R. Kohlmeyer, A. Javadi, B. Pradhan, S. Pilla, K. Setyowati, J. Chen, and S. Gong, *The Journal of Physical Chemistry C*, Vol. 113, p. 17626, 2009.
41. Z. Liu, S.J. Picken, and N.A.M. Besseling, *Macromolecules*, Vol. 47, p. 4531, 2014.
42. Y. Zhang, J. He, and R. Yang, *Polymer Degradation and Stability*, Vol. 125, p. 140, 2016.
43. J. Qiu, X. Lai, W. Fang, H. Li, and X. Zeng, *Polymer Degradation and Stability*, Vol. 144, p. 176, 2017.
44. N. Rezazadeh and J. Rezazadeh, *Journal of Magnetism and Magnetic Materials*, Vol. 475, p. 201, 2019.
45. Z. Yan, X. Liang, I. Cotton, and C. Emersic, *IEEE Transactions on Dielectrics and Electrical Insulation*, Vol. 25, p. 1095, June 2018.
46. R. Hackam, *IEEE Transactions on Dielectrics and Electrical Insulation*, Vol. 6, p. 557, 1999.
47. X. Jun and I.D. Chalmers, *Journal of Physics D: Applied Physics*, Vol. 30, p. 1055, 1997.
48. A. Winter and J. Kindersberger, *IEEE Transactions on Dielectrics and Electrical Insulation*, Vol. 19, 2012.

49. H. Iijima, H. Takahashi, and H. Tagai, Heat curable addition-reactive type silicone rubber composition, US Patent Application 20 150 299 398, assigned to Momentive Performance Materials Japan LLC, Minato-ku, Toyko (JP), October 22, 2015.
50. S. Suzuki, Insulating silicone rubber composition, US Patent 9 056 989, assigned to Sumitomo Riko Company Ltd. (Komaki-shi, JP), June 16, 2015.
51. Y. Hattori and J. Ishizawa, Liquid-encapsulation heat dissipation member, US Patent Application 20 160 278 237, assigned to Polymatech Japan Co., Ltd., Saitama (JP), September 22, 2016.
52. K. Katoh, T. Hirasawa, N. Aze, and T. Endoh, Loop heat pipe wick, loop heat pipe, cooling device, and electronic device, and method for manufacturing porous rubber and method for manufacturing loop heat pipe wick, US Patent Application 20 180 180 360, assigned to Ricoh Co. Ltd., June 28, 2018.
53. V.K.S. Vanita Bhardwaj, Rahul Kumar Gangwar, *Optical Engineering*, Vol. 55, p. 55, 2016.
54. Wikipedia contributors, Mach-Zehnder interferometer — Wikipedia, the free encyclopedia, https://en.wikipedia.org/w/index.php?title=Mach%E2%80%93Zehnder_interferometer&oldid=867316087, 2018. [Online; accessed 18-January-2019].
55. Y. Miyamoto, M. Yoshitake, H. Yoshida, T. Nakata, and K. Hirai, Curable silicone composition and optical semiconductor device, US Patent 9 464 211, assigned to Dow Corning Toray Co., Ltd. (Tokyo, JP), October 11, 2016.
56. Y. Zhang, W. Yu, L. Zhang, J. Yin, J. Wang, and H. Xie, *Journal of Thermal Science and Engineering Applications*, Vol. 10, p. 011014, August 2017.
57. S. Shang, L. Gan, C. Mei, L. Xu, L. Tan, and E. Hu, Wet functionalization of graphene and its applications in rubber composites in S. Yaragalla, R. Mishra, S. Thomas, N. Kalarikkal, and H.J. Maria, eds., *Carbon-Based Nanofillers and Their Rubber Nanocomposites*, chapter 10, pp. 285–322. Elsevier, 2019.
58. L. Wang, P. Liu, L. Li, W. Chen, S. Yang, and H. Zheng, Silicone rubber for led encapsulant, in *2018 15th China International Forum on Solid State Lighting: International Forum on Wide Bandgap Semiconductors China (SSLChina: IFWS)*, pp. 1–3, Oct 2018.
59. M. Takashima, G. Yoshikawa, and T. Kameda, Silicone resin liquid composition, US Patent 9 840 641, assigned to Sumitomo Chemical Co., Ltd. (Tokyo, JP), December 12, 2017.
60. F. Li, Bolt locking structure for waterproof LED lamp, US Patent Application 20 180 224 090, assigned to Feng Li, August 9, 2018.
61. C. Hopmann and M. Röbig, *Progress in Rubber, Plastics & Recycling Technology*, Vol. 33, p. 63, 2017.

62. T. Li, J. Chen, H. Dai, D. Liu, H. Xiang, and Z. Chen, *Journal of Materials Science: Materials in Electronics*, Vol. 26, p. 312, 2015.
63. F.B. Madsen, A.E. Daugaard, S. Hvilsted, and A.L. Skov, *Macromolecular Rapid Communications*, Vol. 37, p. 378, 2016.
64. C. Löwe, X. Zhang, and G. Kovacs, *Advanced Engineering Materials*, Vol. 7, p. 361, 2005.
65. P. Brochu and Q. Pei, *Macromolecular Rapid Communications*, Vol. 31, p. 10, 2010.
66. K. Meijer, M.S. Rosenthal, and R.J. Full, Muscle-like actuators? A comparison between three electroactive polymers, in *Smart Structures and Materials 2001: Electroactive Polymer Actuators and Devices*, Vol. 4329, pp. 7–16. International Society for Optics and Photonics, 2001.
67. R. Pelrine, R.D. Kornbluh, Q. Pei, S. Stanford, S. Oh, J. Eckerle, R.J. Full, M.A. Rosenthal, and K. Meijer, Dielectric elastomer artificial muscle actuators: toward biomimetic motion, in *Smart Structures and Materials 2002: Electroactive Polymer Actuators and Devices (EAPAD)*, Vol. 4695, pp. 126–138. International Society for Optics and Photonics, 2002.
68. E. Delebecq and F. Ganachaud, *ACS Applied Materials & Interfaces*, Vol. 4, p. 3340, 2012.
69. E. Delebecq, N. Hermeline, A. Flers, and F. Ganachaud, *ACS Applied Materials & Interfaces*, Vol. 4, p. 3353, 2012.
70. O. Kretschmann, C. de Santis, and A. Ruppenthal, Curable polyorganosiloxane composition for use as an encapsulant for a solar cell module, US Patent 9 543 460, assigned to Momentive Performance Materials GmbH (Leverkusen, DE), January 10, 2017.
71. C. Terreau, J. de la Croi Habimana, and S. Jenkins, Encapsulation of solar cells, US Patent 8 847 063, assigned to Dow Corning Corporation (Midland, MI), September 30, 2014.
72. M. Gotoh and K. Nakamura, Silicone resin foam and sealing material, US Patent Application 20 160 053 069, assigned to Sekisui Chemical Co., Ltd., Osaka (JP), February 25, 2016.
73. J. Proud, B. Fan, R. Shook, and J. Fernandez, Wearable device made with silicone rubber and electronic components, US Patent 9 542 685, assigned to Fitbit Inc., January 10, 2017.
74. H. Jiang, Q. Yu, H. Li, H. Tao, L. Zhong, and N. Yan, Moisture diffusion property and moisture-induced electrical performances of liquid silicone rubber, in *2018 Condition Monitoring and Diagnosis (CMD)*, pp. 1–5, Perth, Australia, September 2018. IEEE.
75. H. Yu, Y. Huo, M. Esseghir, S. Sengupta, and J. Cogen, In-situ compatibilization of silicone rubber/polyolefin elastomer blends by forming ionomers for cold shrink splice and preparation method thereof, US Patent 10 150 867, assigned to Dow Global Technologies LLC, December 11, 2018.

76. S.V. Davis, R.L. Beckman, and I.S. Olsen, Lubricious silicone cable jackets and cable assemblies formed therefrom, US Patent Application 20 180 023 248, assigned to Creganna Unlimited Co., January 25, 2018.
77. H. Sakakibara and K. Matsunaka, Liquid silicone rubber mixture, and electrophotographic member and fixing device, US Patent Application 20 180 314 172, assigned to Canon Inc., November 1, 2018.
78. S. Nakajima, J. Sugawara, Y. Fukumoto, and N. Onmori, Silicone rubber sponge and rubber-covered roller, US Patent 9 260 606, assigned to Sumitomo Electric Industries, Ltd. (Osaka-shi, Osaka, JP), Sumitomo Electric Fine Polymer, Inc. (Sennan-gun, Osaka, JP), February 16, 2016.
79. K. Ishikawa, S. Katsuki, N. Kusaka, S. Yamazaki, and Y. Matsuzawa, Method of manufacturing heat-fixing rubber roller and heat-fixing rubber roller, US Patent Application 20 170 097 594, assigned to SWCC Showa Cable Systems Co. Ltd., April 6, 2017.
80. T. Kaihara, M. Uno, M. Yamada, and T. Ashibe, Electro-conductive member, process cartridge, electrophotographic apparatus, and addition-curable silicone rubber mixture, US Patent Application 20 160 070 182, assigned to Canon. Inc., March 10, 2016.
81. W.A. Hermann, A. Prilutsky, and V.H. Mehta, Battery pack enclosure with controlled thermal runaway release system, US Patent 8 367 233, assigned to Tesla Motors, Inc. (Palo Alto, CA), February 5, 2013.
82. D.S. Buck, Battery assembly with temperature control device, US Patent Application 20 070 259 258, assigned to EnerDel Inc., November 8, 2007.
83. A. Narbonne, B.T. Turbe, and O. Rabaud, Battery of electrochemical generators comprising a foam as inter-generator filler material, US Patent Application 20 120 003 508, assigned to Accumulateurs Fixes et de Traction SAFT SA, January 5, 2012.
84. V. O'Neil, J. Hanley, M. Kihara, L. Brown, M.J. Watson, and M.P. Timmons, Silicone rubber syntactic foam, US Patent Application 20 180 223 069, assigned to Elkem Silicones USA Corp., August 9, 2018.
85. P. Liu, C. Liu, Y. Huang, W. Wang, D. Fang, Y. Zhang, and Y. Ge, *Journal of Applied Polymer Science*, Vol. 133, 2016.
86. J. Zhu, H. Wang, and Y. Zhu, *Journal of Applied Physics*, Vol. 123, p. 034505, 2018.
87. H. Zhu, F. Wang, L. Wang, and Y. Wang, *Sensors and Actuators A: Physical*, Vol. 271, p. 192, 2018.
88. G. Li, J. Gong, J. Tan, D. Zhu, and W. Jia, Study on mechanical properties of silicone rubber materials used as gaskets in PEM fuel cell environment, in *Proceedings of the ICPNS 16*, pp. 1–6, Seattle

Marriott Waterfront, 2016. 8nd International Conference on Physical and Numerical Simulation of Materials Processing, ICPNS'16.
89. T. Cui, Y. Chao, and J. Van Zee, Stress relaxation behavior of a liquid silicone rubber seal subjected to temperature cycling, in *20th ECS Meeting*, number 16 in ECS Meetings, pp. 1037–1037. The Electrochemical Society, 2011.
90. J. Feng, Q. Zhang, Z. Tu, W. Tu, Z. Wan, M. Pan, and H. Zhang, *Polymer Degradation and Stability*, Vol. 109, p. 122, 2014.
91. M. Schulze, T. Knöri, A. Schneider, and E. Gülzow, *Journal of Power Sources*, Vol. 127, p. 222, 2004. Eighth Ulmer Electrochemische Tage.
92. J. Tan, Y.J. Chao, J.W.V. Zee, and W.K. Lee, *Materials Science and Engineering: A*, Vol. 445-446, p. 669, 2007.
93. J. Tan, Y.J. Chao, X. Li, and J.W. Van Zee, *Journal of Power Sources*, Vol. 172, p. 782 , 2007.
94. J. Tan, Y.J. Chao, M. Yang, C.T. Williams, and J.W. Van Zee, *Journal of Materials Engineering And Performance*, Vol. 17, p. 785, 2008.
95. J. Tan, Y.J. Chao, X. Li, and J.W. Van Zee, *Journal of Fuel Cell Science and Technology*, Vol. 6, p. 41017, August 2009.
96. J. Tan, Y. Chao, M. Yang, W.-K. Lee, and J.V. Zee, *International Journal of Hydrogen Energy*, Vol. 36, p. 1846, 2011. The 3rd Annual Korea-USA Joint Symposium on Hydrogen & Fuel Cell Technologies.
97. T. Cui, Y.J. Chao, and J.W. Van Zee, *Polymer Testing*, Vol. 32, p. 1202, 2013.
98. T. Cui, Y.J. Chao, and J.W. Van Zee, *International Journal of Hydrogen Energy*, Vol. 39, p. 1430, 2014.
99. F. Wu, B. Chen, Y. Yan, Y. Chen, and M. Pan, *Polymers*, Vol. 10, p. 522, 2018.
100. C. Hopmann and M. Röbig, Injection molding of high precision optics for LED applications made of liquid silicone rubber, in *AIP Conference Proceedings*, Vol. 1713, p. 040003, Melville, NY, 2016. American Institute of Physics (AIP).
101. C. Hopmann and M. Röbig, *Journal of Elastomers & Plastics*, Vol. 49, p. 62, 2017.
102. H. Zhu, Z. Dai, and W. Tu, *Materials Science*, Vol. 24, p. 185, 2018.
103. D. Popescu, M. Popescu, C. Cernaianu, and D. Popescu, Regarding the structural analysis of silicone rubber obtained from recycling used composite electric isolators, in *2018 International Conference on Applied and Theoretical Electricity (ICATE)*, pp. 1–6, October 2018.
104. Y. Yang, N. Sun, Z. Wen, P. Cheng, H. Zheng, H. Shao, Y. Xia, C. Chen, H. Lan, X. Xie, C. Zhou, J. Zhong, X. Sun, and S.-T. Lee, *ACS Nano*, Vol. 12, p. 2027, 2018.
105. Wikipedia contributors, Galinstan — Wikipedia, the free encyclopedia, https://en.wikipedia.org/w/index.php?title=Galinstan&oldid=880490951, 2019. [Online; accessed 16-February-2019].

106. P. Zylka, *Smart Materials and Structures*, Vol. 24, p. 065040, May 2015.
107. H.-C. Ma, H.-M. Lee, and H.-P. Yan, Casing covered by silicone rubber and electronic device using same, US Patent 9 298 215, assigned to Hong Fu Jin Precision Industry (Shenzhen) Co., Ltd. (Shenzhen, CN), Hon Hai Precision Industry Co., Ltd. (New Taipei, TW), March 29, 2016.

5

Medical Uses

5.1 Sensors for Medical Application

5.1.1 Piezoresistant Sensor

Piezoresistivity is the change in electrical resistivity under stress or strain. Piezoresistivity may be a property of electrical conductive materials, e.g., electrically conductive rubbers containing conducting nanotubes such as single-wall or multi-wall carbon nanotubes (CNTs) (1). Such materials have been detailed elsewhere (2).

The sensor has an elastic body fabricated from a liquid silicone rubber with a loading of CNTs between 0.5% and 3%, that must be homogeneously distributed therein, with electrodes adhered to or molded with the nanotube composite, and it contains a medical grade non-conductive flexible silicone rubber. The polymer is a platinum-cured liquid silicone rubber composite. It has a hardness of between 10 and 60 *AskerC* and an elastic elongation capability of at least 200%. Asker C is a hardness scale. For example, 60 Asker C is 25 Shore A (3). The properties of silicone depending on CNT loading are shown in Table 5.1.

Figure 5.1 shows an example of a sensor. Figure 5.1 shows a side sectional view of an electrically conductive polymer sensor comprised of nanotubes to confer electrical properties. The sensor is comprised of the cured silicone polymer (or equivalent elastomer or flexible polymer). This is a flexible silicone rubber with carbon nanotube uniformly (essentially homogeneously, within the limits of real physical limits on the use of finite material) dispersed within the polymer at a preferred loading of between 0.5% and 3.0%. On

Table 5.1 Properties of silicone depending on CNT loading (1).

Property	Plain	0.5% CNT	1% CNT	2% CNT
Max.Torque/[$lb - in$]	5.73	6.01	6.61	9.4
TC90	0.7	0.3	0.20	0.5
Hardness, Shore A	38	40	43	43
Tensile/[psi]	365	335	392	449
Elongation/[%]	256	227	230	265
Modulus/[psi]	168	160	97	220
Tear/[ppi]	30.4	28.6	41	52
Sp. Gr.	1.255	1.31	1.31	1.284
Compression se/[%]	2.0	3.2	7.0	12.5
Volume change in silicone oil,% 22h 350°F	34	43	35	35
Electrical resistivity, /[$\Omega\, cm$]	1×10^{13}	5×10^{4}	1×10^{3}	

each end of the material there is an electrical wire (electrode) and connection, which are molded or affixed to the carbon nanotube rubber.

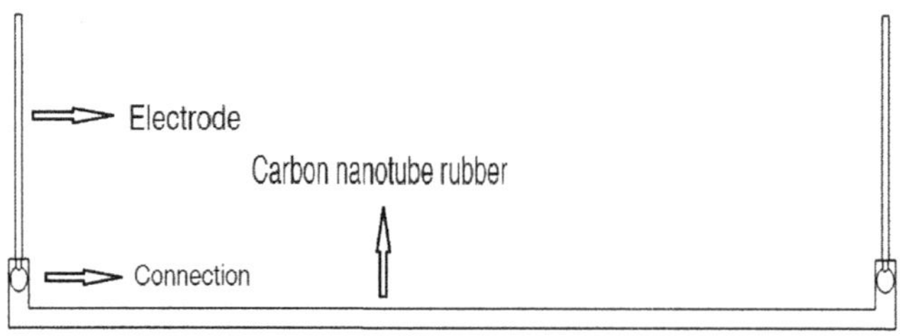

Figure 5.1 Sketch of a sensor (1).

For example, a 0.5% or 1.0% loading of multi-wall CNTs dispersed into a liquid polymerizable to a silicone rubber, changes the resistivity of the original silicone rubber elastomer from 1013 $\Omega\, cm$ to 103 $\Omega\, cm$, with no significant change in the other important properties of the original properties.

Such a sensor can be used to measure elongation or strain of a medical device during insertion or immediately after insertion, or even long after insertion into the patient. Such a sensor can also measure the deformation or load that is placed upon the medical device by the organ or by the body part with which the medical device is in contact. That measurement may be a direct pressure measurement, or arrived at by comparing strain with known degrees of pressure applied perpendicular to the sensor (and using a lookup table). Such a sensor may also be used to measure the amount of pressure that is being applied to a body part by the medical device. The device may also be used to monitor changes over time of the elongation, deformation, strain, load or pressure of an object or body part to which the sensor is affixed (1).

5.1.2 *Pressure Sensor*

A pressure sensing element has been described that has a housing containing a fluid (4). The housing has at least two moveable surfaces in contact with the fluid in the housing; there is a first moveable surface comprising a pressure application surface which, when moved towards, away from, into or out of the fluid, creates a change in fluid pressure and a second moveable surface in contact with the fluid in the housing comprising a membrane which, when deformed by pressure changes in the fluid, alters its electrical resistance.

Figure 5.2 shows a laminated sensor 400 using a membrane 410, with a thickness of about 15 microns, as a component which can sense according to the present technology. The sensor membrane is applied to a flexible contact layer 408, where region 408a has a thickness of about 15 microns and the surrounding flexible contact layer 408b has a thickness of about 50 microns or greater. The flexible contact layer is adhered to a solid support base 402 such that a bubble component 406 is formed. Pressure on the flexible contact layer 408a will deform the membrane sensor 410 to which it is bonded. The deformation changes the electrical resistance of the membrane sensor 410. Electrodes 416 and 418 are used in a manner similar to the electrodes in the previous figures and as elsewhere described herein for applying current or voltage across them and measuring changes upon deformation of the membrane 410.

The membrane has two electrodes attached to the deformable

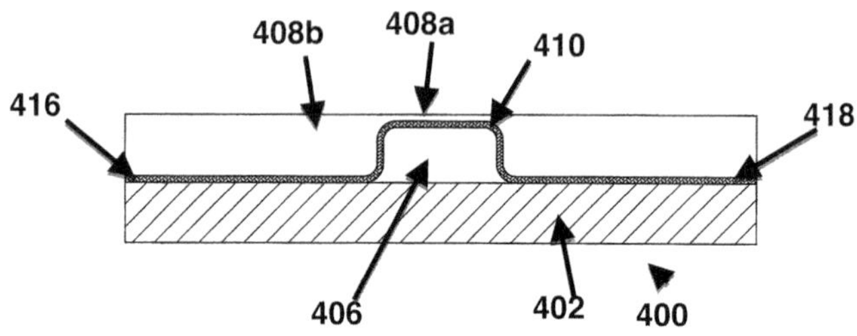

Figure 5.2 Laminated sensor (4).

material in the membrane. The membrane my have an elastomer having conductive particles distributed therein. The elastomer may be a dielectric and the particles are conductive nanoparticles.

Such a sensor can be used to measure the elongation or the strain of a medical device during insertion or immediately after insertion, or even long after insertion into the patient (4).

5.1.3 Flexible Pressure Sensor

A flexible responsive surface of a tool is used on a tool, i.e., robotic or manual, has been described (5). It has a major surface and a sensor attached to and aligned with the major surface of the responsive tool.

The sensor may have a piezoelectric body such as an elastic body containing conductive nanotubes homogeneously distributed therein to form a conductive path and at least two electrodes in electrical connection with the conductive path. Tools used in any surgical or medical treatment that are the subject of medical procedures are particularly useful in methods used in training personnel and evaluating medical techniques and medical personnel.

A medical tool, either functional or itself a partial or complete replica, e.g., needles, scalpels, blades, catheters, syringes, stents, saws, lasers, implants, prostheses, struts, supports, tongs, medical pliers, pacemakers or defibrillators, may carry at least one of the electrodes, such that a circuit to provide a signal is completed only by introduction of the medical tool into the responsive model environment. The overall signals from the system may be effectively neutral until the medical tool is brought into the procedural environment.

Upon proximity to the sensors in the responsive models, contact with a surface that alters the position, tension, pressure, dimensions and/or distribution of or within the surface or body of the sensor, e.g., moisture indicators/sensors and any other physical events that cause altered signals or responses from the sensors, the electrode in the medical tool acts as at least one of the required two electrodes in the responsive model system (5).

5.1.4 *Intraocular Pressure Sensor*

Intraocular pressure is a primary indicator of glaucoma, but measurements from a single visit to a clinic miss the peak intraocular pressure that may occur at night during sleep. A soft chipless contact lens sensor that allows the intraocular pressure to be monitored throughout the day and at night has been developed (6).

A resonance circuit composed of a thin film capacitor coupled with a sensing coil that can sense the corneal curvature deformation was designed, fabricated and embedded into a soft contact lens. The resonance frequency of the sensor is designed to vary with the lens curvature as it changes with the intraocular pressure. The resonance circuit is then sandwiched between two medical grade silicone rubbers, NuSil MED-6015, liquid silicone elastomer, NuSil Technology LLC. The contact lens sensor with sensing elements was embedded in a silicone rubber contact lens (6).

The frequency responses and the ability of the sensor to track intraocular pressure cycles were tested using a silicone rubber model eye. The results showed that the sensor has excellent linearity with a frequency response of $\sim 8\ kHz\, mmHg^{-1}$, and the sensor can accurately track fluctuating intraocular pressure. These results indicated that the chipless contact lens sensor can potentially be used to monitor the intraocular pressure to improve the diagnosis accuracy and the treatment of glaucoma (6).

5.1.5 *Rod Template*

One type of spinal surgery is spinal fusion surgery (7). In spinal fusion surgery, vertebrae are joined together with spinal rods. Spinal fusion surgery is used for spinal stenosis, spinal disc herniation, scoliosis, spinal cord trauma, spondylolisthesis or other spinal disorders.

A rod template may be used in spinal fusion surgery. The rod template is used for determining a curvature and a length of the spinal rods that are used for joining the vertebrae.

A rod template is used for determining the curvature and the length of a spinal rod used for joining vertebrae in a spine fusion surgery in advance of the joining of the vertebrae. The rod template consists of (7):

1. A transparent outer tube made of silicone rubber, with at least one end having an opening, and a hollow portion defined by an inner surface of the transparent outer tube that communicates with the opening,
2. A core made of metal and including scale marks on a surface thereof, the core being disposed in the hollow portion of the transparent outer tube with a gap between the inner surface of the transparent outer tube and the core, and
3. A transparent filler made of liquid silicone rubber that completely fills the gap between the inner surface of the transparent outer tube and the core, wherein the hollow portion of the transparent outer tube has a length that is equal to or greater than a length of the core.

Specifically, the silicone rubber is a material that is used for artificial blood vessels (7). The core may be made of a shape-memory alloy such as a nickel titanium alloy. Because the shape-memory alloy recovers its original shape when the shape-memory alloy is heated to a shape recovery temperature, it is preferable to use the shape-memory alloy for a rod template that is repeatedly used.

The filler is made from liquid silicone rubber. The liquid silicone rubber is a kind of rubber having a property to turn into an elastomer with crosslinks at a temperature in a range from room temperature to about 60°C, which is a relatively low temperature range. Before crosslinking occurs, the liquid silicone rubber has a low viscosity and shows flowability. There are room-temperature curing type and thermosetting type liquid silicone rubbers classified according to their response temperatures.

The length of the spinal rods is determined by a measurement with the scale marks. The outer tube is used for restricting small metal pieces from spreading if cracking of the surface of the core occurs when the rod template is bent and such small metal pieces are produced from the cracking (7).

5.1.6 *Cupping Appliance Device*

A cupping appliance is a traditional Chinese medical healthcare appliance, which has existed since ancient times and has astounding effects (8). A traditional cupping appliance normally is made with solid material such as ceramics, bamboo cone, or glass. A suction cupping appliance made with organic glass also has been applied. The cupping appliance made with solid material normally needs ignition or gas sucking so as to allow the internal cavity to form a definite negative pressure, such that suction may be performed on a specific part of a human body. Cupping with ignition may easily cause burns while cupping with suction if it is not sufficiently convenient to operate. Furthermore, the cupping appliance made with solid material may easily fall off because the skin swells up after cupping. It is difficult to cup the slim muscles and curvy parts, such as the joints, spine, etc. In addition, it is not convenient to carry the cupping appliance.

In recent years, a novel cupping appliance that is made with rubber or silicone rubber has been introduced. Compared with the cupping appliance made with hard material, the cupping appliance made with rubber or silicone rubber material has the following characteristics (8):

1. No need to ignite, will not burn a human, easy to learn how to operate, and easy to operate,
2. Large cupping force, firm cupping, and the cupping sucking force may be easily adjusted,
3. Since the cup opening is soft and may change shape, joints, spines, and other parts with curvy surfaces may also be cupped,
4. No limitation on physical status of the human, may perform cupping while sitting, standing or even walking,
5. Not easily broken, convenient to carry.

Since the cupping appliance made with rubber or silicone rubber possesses the above-mentioned beneficial characteristics, and is favored by medical practitioners and patients.

However, the rubber material applied in the current rubber cupping appliance is not transparent. In addition, the silicone rubber cupping appliance also may only be semitransparent due to limitations in the manufacturing process. This would affect observations made by medical practitioners towards the concerned skin parts of the patients. This is also the main deficiency for rubber elastic cupping.

The conventional manufacturing technique of the rubber or silicone rubber cupping appliance uses a molding technology. The general process for the molding technology is as follows:

After the solid silicone rubber material is mixed through the mixing mill, a mixed film is cut using a cutting machine, into a size and width suitable for filling a mold cavity. The rubber with weight slightly greater than that of the product is weighted out. Then it is placed inside the mold cavity. The mold is installed on a pressure-forming machine, then the machine is turned on. The molding is formed under a certain temperature maintained for a duration of time. Finally, the cupping appliance is completed after demolding. The molding possesses the following deficiencies: First, the production procedure includes rubber mixing, material mixing, material cutting, material weighting, material placing, mold installing, vulcanization molding, demolding, and trimming. The production procedure is complicated, time-consuming, and has low production efficiency. Second, the molding method is rough and not meticulous, and it is difficult to avoid bad product quality caused by human operation in the process of molding. Third, the demolding is difficult. During the process of demolding, the cupping appliance may easily be damaged. The damage rate is high. At the same time, in the process of demolding, the die needs to be dismounted, which wastes time and energy, and increases the labor force and lengthens the production time.

In order to solve the above-mentioned problems, another method and a device for producing a silicone rubber cupping appliance has been developed. Using liquid silicone rubber as the raw material and applying injection of silicone rubber vulcanization molding technology, a silicone rubber cupping appliance possessing high elasticity, high transparency, and complex shapes may be produced. Furthermore, the procedure is simple, the whole process is under automatic operation, the productivity rate is high, the demolding is simple and fast, and the demolding process does not require the device to be dismounted for cleaning.

An injection vulcanization molding device specifically used to produce a silicone rubber cupping appliance is formed, so as to enable the manufacture of a silicone rubber cupping appliance with high elasticity, high transparency, and complex shapes using liquid silicone rubber as the material and through the injection vulcanization molding technique.

The device is shown in Figure 5.3.

The method for making the silicone rubber cupping appliance includes the following steps:

1. Heating the molding zone and cooling the channel that input where the liquid silicone rubber material flows pass, such as the nozzle 12 (namely, the channel which is inputted with liquid silicone rubber material), or cooling the zone surrounding such channel with the liquid silicone rubber material. If the channel where the liquid silicone rubber material may flow pass includes the cold flowing channel, then the cool flowing channel must also be cooled, so as to prevent the liquid silicone rubber material from heating up and vulcanization molding in the transport process.
   Such liquid silicone rubber material may include a liquid additive that must be added into the liquid silicone rubber. Therefore, the liquid silicone rubber material including the liquid additive may also be named as the liquid silicone rubber material. The molding zone includes a slide 7, an internal mold core 4 and a mold cavity 1 formed from the slide 7 and internal mold core 4, as illustrated in Figure 5.3. The method of heating up the molding zone may be performed through the heating device installed at the molding zone, such as the heating pipe or heating panel or heating film which supplies the energy. Energy may also be supplied through external heating devices. The energy supplied should to the greatest extent allow each part of the molding zone to maintain the same or similar temperature. That way, each part of the

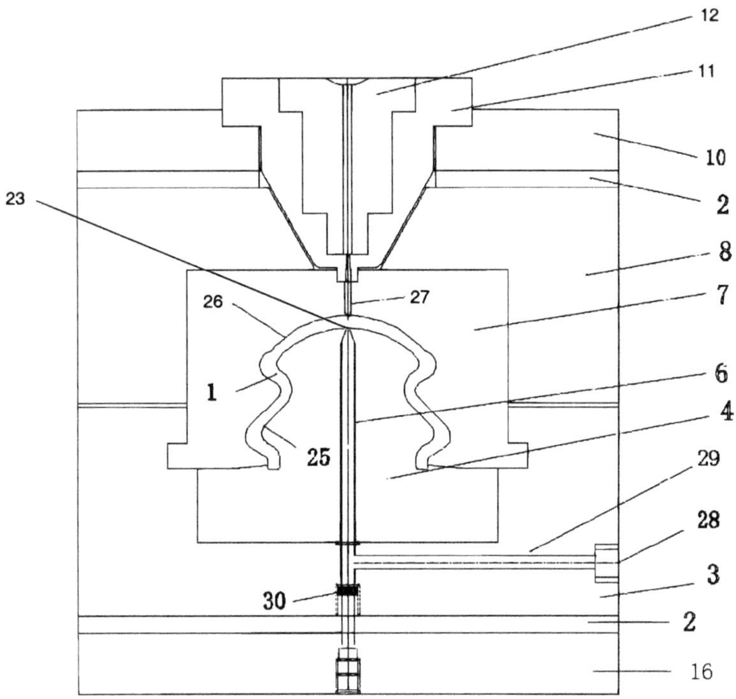

1	Mold cavity	2	Insulation board
3	Rear mold board	4	Internal mold core
6	Blowing-demolding unit	7	Slide
8	Front mold board	10	Panel
11	Nozzle cover	12	Nozzle
16	Blow needle fastening board	23	Blow duct opening
25	Internal mold surface	26	External mold surface
27	Hot flowing channel	28	Gas entry opening
29	Gas entry channel	30	Sealing pad

Figure 5.3 Injection vulcanization molding device (8).

silicone rubber cupping appliance in the mold cavity 1 may possess a balanced quality.

2. Injecting the liquid silicone rubber material into the mold cavity 1, such as the liquid silicone rubber material may be injected into the mold cavity 1 through the nozzle 12, and/or cold flowing channel, and/or hot flowing channel 27.
3. Ceasing the addition of materials, and preserving heat for a duration of time, so as to allow the liquid silicone rubber material to vulcanization solidifying and molding.
4. Separating the separable slide 7, at the same time opening the blow duct opening. The blowing-demolding unit 6 blows the silicone rubber cupping appliance in the mold cavity 1 causing it to be demolded. If the blowing-demolding unit 6 is mainly formed by the blow duct opening 23, blow duct, then the blow duct opening 23 and the mold cavity 1 are connected through the retreating of the blow needle, then the mold cavity 1 is blown with high pressure gas by the high pressure gas source through the blow duct, so as to allow the silicone rubber cupping appliance to demold after molding. In particular, step 4 may preferably include the following specific steps:
5. Separating the separable slide 7 at the two sides along a direction lying vertical or approximately vertical to the mold cavity 1, causing an external mold surface 26 of the silicone rubber cupping appliance in the mold cavity 1 to be exposed to the air. Obviously, it may also be a plurality of separable slides provided along the axial direction of the mold cavity 1 lying distal from the mold cavity 1 along the axial direction of the mold cavity 1. At the same time, the external mold surface 26 of the silicone rubber cupping appliance in the mold cavity 1 may be exposed to air.
6. Opening the blow duct opening 23 in the blowing-demolding unit 6, causing the blowing-demolding unit 6, such as the blow duct, to connect to the mold cavity 1. That is, the internal mold core 4 and the silicone rubber cupping appliance under vulcanization solidifying and molding inside the mold cavity 1 move relatively together in an opposite direction relative to the blow needle. A gap is formed between the blow needle and the internal mold surface 25 of the mold

cavity 1.

7. Off blowing-demolding: The high pressure gas source blows out high pressure gas from the internal mold surface 25 of the mold cavity 1 through the blow duct 21 and the blow duct opening 23 of the blowing-demolding unit 6, causing the solidified silicone rubber cupping appliance to detach from the internal mold surface 25. The pressure of the high pressure gas is preferably at 6–7.5 *atm*, which may speedily and smoothly blow out of the molded silicone rubber cupping appliance.

5.2 Materials for Medical Instruments and Uses

5.2.1 *Wound Regeneration*

A porous structure is critically important for wound dressing or tissue engineering scaffolds (9). However, the influence of the pore sizes on cell proliferation, tissue regeneration and the underlying mechanism were not investigated in the past.

Silicone rubber membranes with different pore sizes were prepared using certain constituents of liquid silicone rubber precursor/liquid paraffin/hexane. It was found that the pore size had a significant impact on cell proliferation and wound healing. The CCK8 analysis revealed that the membrane with a certain pore size (110.47 μm, middle pore membrane) was suitable for cell proliferation compared with the membranes with other pore sizes 218.03 μm, large pore membrane, 5.27 μm, small pore membrane and nonporous membrane (9).

Further studies demonstrated that the middle pore membrane promoted cell proliferation via activating the Wnt/β-catenin signaling pathway (10). In addition, wound healing experiments showed that 7 *d* post-wounding, the rate of wound healing was 89.25% with the middle pore membrane, which was significantly higher than with large pore membrane, small pore membrane, or nonporous membrane (9). The *in-vivo* data indicated that wound healing was accelerated by treatment with a silicone rubber membrane with a pore size of 110.47 μm.

The results of the study strongly suggested that different pore structures might affect the cell proliferation and wound healing and that a silicone rubber membrane with a specific pore size could potentially be used as a promising wound dressing (9).

5.2.2 *Prostate Brachytherapy*

A transparent material with similar structural characteristics and mechanical properties to human tissue is favorable for prostate brachytherapy (11).

A transparent poly(vinyl alcohol) (PVA) hydrogel and silicone rubber were developed as suitable substitutions for human soft tissue. Segmentation and 3D reconstruction of medical image were performed to manufacture the mold of organ groups through rapid prototyping technology. Microstructure observation, force test and a CCD deformation test have been conducted to investigate the structure and mechanical properties of the PVA hydrogel used in organ group mockup.

Cleidocranial dysostosis (CCD), also known as cleidocranial dysplasia, is a birth defect that mostly affects the bones and teeth (12). The collarbones are typically either poorly developed or absent, which allows the shoulders to be brought close together.

Scanning electron microscopy (SEM) results showed that a PVA hydrogel consisting of 3 *g* PVA, 17 *g* de-ionized water, 80 *g* dimethylsulfoxide, 4 *g* NaCl, 1.5 *g* NaOH, 3 *g* epichlorohydrin and 7 freeze/thaw cycles results in a similar microstructure to human prostate tissue.

Through the insertion force comparison between organ group mockup and clinical prostate brachytherapy, PVA hydrogel and silicone rubber were found to have the same mechanical properties as prostate tissue and muscle. CCD deformation test results showed that insertion force suffers a sharp decrease and a relaxation of tissue deformation appears when a needle punctures the capsule of the prostate model. The results show that organ group mockup consisting of PVA hydrogel, silicone rubber, membrane, and agarose satisfies the needs of prostate brachytherapy simulation in general and can be used to mimic the soft tissues in pelvic structure (11).

5.2.3 *Breast Implants*

Silicone breast implants are implanted under the breast tissue or under the chest muscle for breast augmentation or for the purpose of breast reconstruction, for example, following breast cancer or benign breast diseases (13). Breast implants possess an outer shell based on crosslinked silicone that is normally filled with a silicone gel consisting of a weakly crosslinked silicone and uncrosslinked silicone. Breast implants are available in many different sizes and have either smooth or textured shells. Materials for silicone breast implants and their properties are described in the literature (14,15).

The implants are made of poly(dimethyl siloxane) and consist of a highly crosslinked silicone shell with a lesser crosslinked silicone cell inside.

It is estimated that 15.000 to 20.000 breast augmentations are performed annually in Germany and several million worldwide (16).

A possible aging process affecting a silicone implant is the infiltration of body compounds into the shell. Lipids can trigger the degradation of the elastomer by causing swelling of the polymer network, a process which leads to progressive weakening of the silicone elastomer envelope, especially with regard to its tensile strength (17,18). Via liquid-state nuclear magnetic resonance experiments, Birkefeld *et al.* detected absorbed phospholipids in both the shell and the gel (18).

Another factor which is responsible for deterioration in the mechanical properties of the shell could be the shell-gel coupling. The diffusion of non-crosslinked silicone from the gel into the shell causes swelling of the envelope, and this time-dependent phenomenon results in a drop in the mechanical properties of the shell, with an asymptotic increase in shell swelling and a decrease in shell strength properties (19,20)

To obtain a better knowledge of the reason for and the time of a possible failure of the implant shell, it is important to acquire information regarding the stability and durability of the material used and the associated failure criteria (16).

Therefore, the mechanical properties of silicone elastomers have been investigated with regard to breast implants. Sample strips of the elastomer of various thicknesses were prepared and exposed to a cyclic bending load for a number of load cycles. Tensile specimens were punched out according to a defined pattern from the preloaded sample strips and tested with regard to their tensile properties, particularly the influence of cyclic loading on the stress at break.

However, as quality of a silicone implant depends not only on the strength of the material, but also on the spatial variation in the mechanical properties, the quality of the silicone samples was also assessed in terms of the characteristic stress at break and the Weibull modulus, applying Weibull statistics to the tensile test results.

The study revealed that the characteristic stress at break remained constant with increasing number of load cycles, but that the Weibull modulus decreased exponentially during the first 10.000 cycles to a critical level (16).

5.2.4 *Implant with Reinforcing Fibers*

Improved materials have been developed with an enhanced stability and compatibility with the surrounding tissue (13).

It has been a problem to determine and optimize the materials with respect to stability, biocompatibility, and safety. The implant has to remain stable and functional for a long time, over years or even decades.

Studies showed that damages of such implants with time, especially rupture and cracks, are a tremendous practical problem. It was found that long-term *in-vivo* stability of the implants could not simply be correlated to mechanical stability determined theoretically. It is difficult to provide implants having long-term stability, which at the same time have a good biocompatibility and elasticity in the human body. Moreover, it is difficult to monitor and evaluate such materials which are not accessible *in vivo*. Overall, the mechanisms for adjusting long-term stability and avoiding damages over time were not sufficiently understood.

Breast implants in which a silicone matrix is reinforced with poly(urethane) (PU) fibers have been described (21). However, the compatibility of PU with polysiloxanes is generally not adequate. The compatibility of the components of such composite materials is highly important after implantation, because even minor segregation of the components will result in severe problems after years or decades and, ultimately, the overall structure can rupture or collapse (13).

Breast implants have been developed which overcome the problems described above (13). An implant is provided which has an improved stability. Especially, the long-term stability will be advantageous. The susceptibility to damage, especially rupture and cracks, should be low. Furthermore, the biocompatibility and safety should be adequate.

An outer shell for a breast implant is used, where the outer shell is a fiber-reinforced silicone containing a silicone matrix and fibers embedded in the silicone matrix, wherein the fibers are a comb polymer having a base polymer and side chains. The base polymer is an organic polymer and the side chains are polysiloxanes. Furthermore, a breast implant is at least partly elastomeric and the elasticity is adapted to the female breast.

The fibers are embedded in the silicone matrix. The fibers are a comb polymer having a base polymer and side chains, wherein the base polymer is an organic polymer and the side chains are polysiloxanes. Examples of the polymers are poly(methyl methacrylate) grafted poly(dimethyl siloxane) copolymers.

The used fibers are oriented. This means that the fibers are not laid in random orientation, but are aligned in a special direction. Preferably, the fibers are aligned essentially in parallel. For example, electrospun fibers can be oriented by charging the support on which the fibers are spun, such that it determines the deposition direction of the spun or emerging fibers. The fibers are preferably oriented such that they stabilize weak regions of the implant and increase the resilience of the implant. Preferably, the oriented fibers stabilize tangential regions of the implant. Preferably, the fibers are oriented perpendicular to the equatorial region and/or to the equatorial plane. Methods for aligning fibers by electrospinning are known in the art (13).

5.2.5 *Hair Implants*

A variety of hair replacement techniques and methods exist, such as hair pieces and toupees, hair weaves and extensions, hair implants and hair transplant surgery (22).

Medical pharmacological therapy, such as Finasteride and Minoxidil, can prevent or stimulate hair growth directly (23–28). The drugs Finasteride and Minoxidil are shown in Figure 5.4.

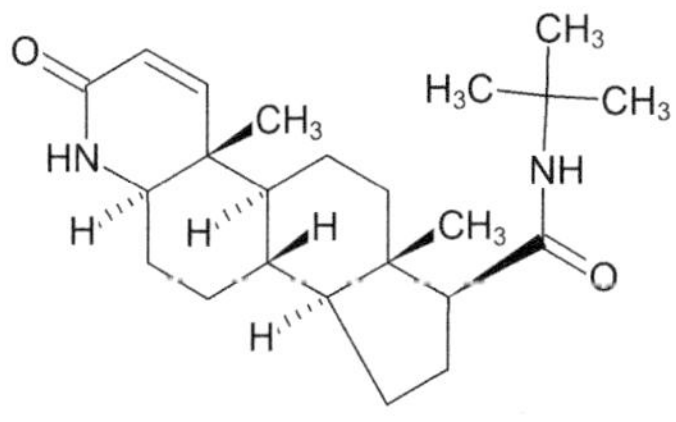

Finasteride

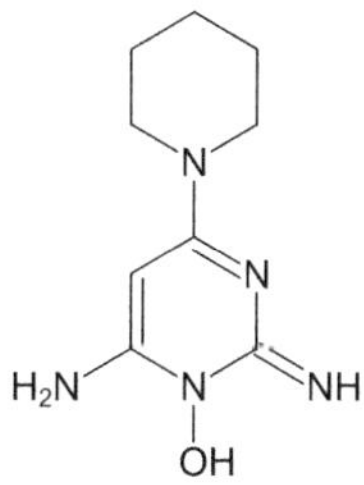

Minoxidil

Figure 5.4 Drugs.

These medications not only perform poorly regarding hair regrowth, but are also not benign treatments having no risks or side effects. Finasteride, for example, can not only cause a loss in libido but it can increase the risk of developing a more aggressive type of prostate cancer. Minoxidil, a topical medication, is problematic because it can grow hair in other parts of the body such as the face arms legs and chest, and this is a side effect that many men and especially women find disturbing. In addition, Minoxidil can lower blood pressure and cause fainting (22).

A hair implant has been developed consisting of (22):

1. At least two strands including at least one of mammalian hair and synthetic hair, and
2. An anchor, which:
   (a) includes silicone,
   (b) is configured for subcutaneous implantation,
   (c) includes a fracture line configured to facilitate fracturing of the anchor along the fracture line for ease of removal of the implant after subcutaneous implantation, and
   (d) is configured to provide a scaffold for collagen growth after subcutaneous implantation.

Here at least one of the two strands is joined to the anchor on one side of the fracture line and at least one of the two strands is joined to the anchor on an opposite side of the fracture line such that each fragment formed by fracturing the implant includes at least one of the two strands.

A second aspect is a hair restoration method consisting of (22):

1. Inserting a needle into the skin to form an incision,
2. Inserting an implant in the incision such that 0.1–2 *mm* of a silicone coating on a proximal end of each of the at least two strands remains:
   (a) outside the skin,
   (b) above the epidermis,
   (c) under the skin, or
   (d) below the epidermis, and

   applying an adhesive to the incision.

In certain embodiments of the method, the incision is made to a depth of 2–8 *mm*. The adhesive can be cyanoacrylate.

A third method for manufacturing the implant consists of (22): providing a mold which includes at least one cavity for forming the anchor; filling at least one cavity with a silicone liquid, coating 2–10 *mm* of a proximal end of at least one strand with a silicone coating, submersing in the silicone liquid the proximal end of at least one strand to a depth such that 0.1–2 *mm* of the silicone coating remains outside of the silicone liquid in the cavity, curing the silicone liquid to provide a solid product, removing the solid product from the mold, and sterilizing the solid product so as to provide the implant.

Medical grade silicone rubber and similar materials have been used for cosmetic and medical use for decades as major components of implantable medical devices (22). These implants have been proven safe for permanent implantation into the body. Even though safe, it is recognized that there is the possibility of the need for future replacement of such implants over time. Implant replacement is indicated, for example, in the case of long-standing, malformed or ruptured silicone breast implants (and other cosmetic implants), when severe contracture, pain, psychological issues, or an aggressive foreign body reaction with collagen encapsulation of the medical device results (29–35).

5.2.6 *Nasal Implants*

An implant for augmentation rhinoplasty, an additive plastic nose surgery, was designed to suit the face morphology of patients of the Indian subcontinent and to reduce the cost substantially by incorporating innovative design and manufacturing practices, thus overcoming the problems in the design of existing implants and the high costs.

Liquid silicone rubber (LSR) was chosen as raw material because of the ready availability and its ability to be molded into the desired shape. An LSR-25A with a Shore Hardness 25A was selected. The implants were designed using Solid Edge software (36).

5.2.7 *Injectable Implants*

The usefulness of addition type liquid silicone rubber as injectable implant after evisceration to maintain the eyeball volume were investigated in an animal experiment (37).

Twelve adult New Zealand white rabbits were tested. One eye of each rabbit was randomly selected for evisceration with the fellow eye as control. The material was injected to fill the eyeball socket after evisceration. *In-vivo* observation was carried out and photographs were taken up to 24 weeks post-op.

Two rabbits were sacrificed respectively at post-operative week 1, 2, 4, 8, 12, and 24. After enucleation, the vertical, horizontal and sagittal diameters of the experimental eyeballs were measured and compared with the control eyes. Histopathological studies were performed to evaluate signs of inflammation (37).

The cornea remained clear throughout the observation period, despite mild epithelial edema and neovascularization. Compared to the control eyes, the experimental eyes were significantly smaller in vertical diameter (17.00 ± 1.17 vs. 17.54 ± 1.11 *mm*, $P<0.001$), but larger in sagittal diameter (16.85 ± 1.48 vs. 16.40 ± 1.38 *mm*, $P = 0.008$), and had no significant difference in horizontal diameter (17.49 ± 1.53 vs. 17.64 ± 1.21 *mm*, $P = 0.34$).

Postoperative inflammation was observed at one week after surgery, which peaked at 2–3 weeks, then regressed gradually. At week 12 and week 24, most of the inflammatory cells disappeared with some residual plasma cells and eosinophils. Eosinophils are a variety of white blood cells and one of the immune system components responsible for combating multicellular parasites and certain infections in vertebrates. Along with mast cells and basophils, they also control mechanisms associated with allergy and asthma (38).

An injectable addition type silicon rubber may be a good choice for ocular implantation after evisceration, maintaining eyeball volume and cosmetically satisfactory when compared to the fellow eye. Spontaneous regression of inflammation implied good biocompatibility for at least 24 weeks (37).

In recent years, deoxyribonucleic acid (DNA) molecules have been synthesized to include any useful set of information as encoded via the ordinary DNA building blocks A, T, C, and G (39). George Church at Harvard and other labs around the world, have recently shown that this well-known genetic code, which is used in nature to encode the core information of life, can instead be used in a biotech context to encode ordinary alphanumeric information (40).

The nucleic acid building blocks A, T, C, and G from which DNA is made comprise a $N = 4$ (quaternary) digital code, much as bars on an ordinary barcode is comprised of an $N = 2$ (binary) digital code.

Based on that simple idea to convert DNA information (A, T, C, and G) into an equivalent alphanumeric code, it has been shown that DNA can be a medium for extremely high density data storage. A number of standard *DNA alphabets* have been discussed, which relate each of the possible 64 base triplets (e.g., A,T,C) to a single letter or number, thus allowing any alphanumeric symbol to be converted to an equivalent triplet-based DNA code.

A stenographic method for concealing coded messages in DNA was already shown in 2001 (41). This method consists of concealing a DNA encoded message within a genomic DNA sample followed by further concealment of the DNA sample to a microdot.

A method of incorporating DNA into an elastomeric material has been developed. The method includes the steps of (39):

- Treating DNA with an additive to form a treated DNA, wherein the additive is a polyol, diol, glycol, starch, or pyrrolidone,
- Mixing the treated DNA with an elastomeric material to form a mixture,

- Molding the mixture, and
- Curing the mixture.

The mixture may be cured at temperatures equal to or greater than about 150°C. Curing can take place by press curing and post curing. Types of molding include press molding, injection molding, and blow molding (39).

5.2.8 *3D Printing of Medical Implants*

The diagnosis and treatment of patients suffering from neurological diseases with patient-individualized silicone rubber-based implants is one of the most promising and challenging approaches to improve treatment results (42).

Medical additive manufacturing techniques were developed for fabrication of such implants, but they did not achieve the required printing resolution because of an intensive droplet spreading of the initially liquid silicone rubber, poly(dimethyl siloxane), on the printing substrate.

While empirical optimization approaches for the droplet spreading are costly and time-consuming, a mathematical optimization approach has been assessed to calculate the optimal printing parameters for minimal droplet spreading. Since the viscosity profile of thermal curing silicone rubber is the main reason for the droplet spreading.

A rheology model was implemented for the calculation of the optimal heat curing parameters. A Dual Arrhenius equation was used to correlate the temperature-time-profile of the curing process with the curing-related viscosity rise and the temperature-related viscosity fall of the liquid silicone rubber used therein. Two commonly used silicone rubbers were characterized with a rheometer at different isothermal and anisothermal curing profiles.

A correlation between the calculated and the measured viscosity profiles was observed. This resulted in the ability to optimize the curing process parameters to the rheological behavior of the silicone rubber used (42).

5.2.9 Voice Prostheses

Silicone rubber is widely used as medical implants or devices due to its inactivity to the immune system, good sterilization opportunities, anti-adhesive and stable properties during long-term residence in the human body, and minimum negative tissue response (43,44).

The prosthesis was made of implant grade silicone rubber MED-4750 (45,46) from NuSil Technology (Carpinteria, CA, USA) (47). It is a two-part, high tear strength silicone elastomer, which consists of dimethyl and methylvinyl siloxane copolymers and reinforcing silica. To process prostheses using MED-4750, Part B was softened first on a cooled two-roll mill, and then, Part A was softened. An equal portion by weight of softened Part B was added, and next, these components were cross-blended until thoroughly mixed. The temperature of the blended material is kept as low as possible to give maximum table life. The curing process was carried out at 116°C for 10 *min*.

The influence of a smooth surface molding technique of silicone rubber indwelling voice prostheses on *in-vitro* biofilm formation and analysis of the clinical *in-situ* lifetime has been evaluated (47).

The biofilm formation on smooth and Groningen ultra low resistance prostheses was studied in an artificial throat model. The clinical lifetime of smooth voice prostheses was compared to the previous lifetime of URL by counting the number of replacements in a consecutive 6-month period in the same patient. Eleven laryngectomized patients in follow-up testing required frequent replacement of their voice prostheses.

The use of a smoother mold and less viscous silicone rubber yielded a decrease in surface roughness from 46 to 8 *nm* and was accompanied by a 40% reduction in the prevalence of bacteria and yeast in *in vitro* formed biofilms. Clinically, the lifetime was significantly ($P<0.005$) increased by a factor of 2.1.

The study suggested that the choice of material and in particular its surface finishing may be determining factors with respect to the clinical lifetime of silicone rubber implants and devices failing due to biofilm formation (47).

Silicone rubber has a hydrophobicity in the bio-abhesive range, which constitutes another reason for its widespread use as a biomaterial, because infection is common in the clinical application of biomaterial implants, and almost inevitably leads to removal of the implant (48).

This is also the case for voice prostheses, despite the fact that they are made out of bio-abhesive silicone rubber. The positioning of the valve in the unsterile environment of the esophagus causes rapid biofilm formation and malfunctioning of the valve. The biofilm on voice prostheses consists of a mixed biofilm of bacteria and yeast.

An important aspect of a biomaterial surface influencing biofilm formation is its roughness. The surface roughness of an end product is influenced by the material used and the surface of the mold, which is influenced by the polishing of the mold (48).

5.2.10 *Implantable Medical Leads*

Implantable medical leads are devices that deliver electrical signals to implantable medical devices (49). Exemplary implantable devices are cardiac rhythm management systems, e.g., pacemakers, defibrillators, and cardiac resynchronization therapy devices, and neurostimulation systems, e.g., spinal cord stimulation systems. For cardiac rhythm management systems, medical leads are typically advanced intravascularly to an implant location within or on a patient's heart. For neurostimulation systems, such leads are typically positioned beneath the skin, in vessels located in the neck or limbs, in the pectoral region, in the epidural space of the spinal cord, or intramuscularly.

Implantable leads typically include a flexible conductor surrounded by an insulating tube or shaft that extends from an electrode at the distal end to a connector terminal at the proximal end.

A flexible conductor surrounded by an insulating tube or shaft has been developed that extends from an electrode at the distal end to a connector terminal at the proximal end. Many leads incorporate multiple connectors extending from an electrical contact on a connector terminal to an electrode on a distal end of the lead body. When the connector terminal is coupled to an implantable device, and the device and lead are implanted in a patient, certain stresses or strains may develop in portions of the lead body or conductors near the terminal connector, or regions of a lead that experience bending.

A method for the fabrication consists of injecting a silicone rubber having a durometer of between about Shore 60 D and about Shore 40 A over a core pin to form a lead body, wherein the core pin is made of a shape-memory metal, the core pin having a first generally sinusoidal configuration and a second generally linear configuration.

The core pins can be made of a shape-memory alloy having an undulating austenitic shape and a straight martensitic shape. In these embodiments, the core pin can be used in an overmolding process, then placed in a cold temperature environment (e.g., dipped in cold water) to straighten the core pin to facilitate removal (49).

5.2.11 *Cochlear Electrode Array*

In human hearing, hair cells in the cochlea respond to sound waves and produce corresponding auditory nerve impulses (50). These nerve impulses are then conducted to the brain and perceived as sound.

A damage of the hair cells results in a loss of hearing, because the sound energy, which is received by the cochlea is not transduced into auditory nerve impulses. This type of hearing loss is called *sensorineural deafness*. To overcome sensorineural deafness, cochlear implant systems, or cochlear prostheses, have been developed (50).

A cochlear electrode array is a thin, elongated, flexible carrier containing several longitudinally disposed and separately connected stimulating electrode contacts. The electrode array may be constructed out of biocompatible silicone, platinum-iridium wires, and platinum electrodes. This gives the distal portion of the lead the flexibility to curve around the helical interior of the cochlea.

To place the electrode array into the cochlea, the electrode array may be inserted through a cochleostomy, or via a surgical opening made in the round window of the cochlea. The electrode array is inserted through the opening into the scala tympani, one of the three parallel ducts that make up the spiral-shaped cochlea. The electrode array is typically inserted into the scala tympani duct in the cochlea to a depth of about 13 to 30 *mm*.

The cochlear electrode array delivers electrical current into the fluids and tissues immediately surrounding the individual electrode contacts to create transient potential gradients that, if sufficiently strong, cause the nearby auditory nerve fibers to generate action potentials. The auditory nerve fibers branch from cell bodies located in the spiral ganglion, which lies in the modiolus, adjacent to the inside wall of the scala tympani. The density of electrical current flowing through volume conductors, such as tissues and fluids, tends to be highest near the electrode contact that is the source of such current. Consequently, stimulation at one contact site tends to selectively activate those spiral ganglion cells and their auditory nerve fibers that are closest to that contact site.

A method of forming a cochlear electrode array includes (50):

1. Forming a first foil into a tethered set of electrode pads.
2. The tethered set of electrode pads is mechanically attached to a sacrificial substrate and the tethers are cut.
3. A second foil of flexible electrically conductive material is formed into a tethered set of winged tabs.
4. The tethered set of winged tabs is then attached to the electrode pads and the tethers cut to release the winged tabs.
5. The appropriate wires can then be connected to the corresponding electrode assemblies.
6. The wire is attached to the winged tabs by folding a tab over the wire and welding the tab in place to hold the wire.
7. The remaining wires are shaped into the desired wire bundle geometry by folding up the wings.
8. The sacrificial substrate can then be removed and the electrode array encapsulated in a flexible polymer body.
9. In embodiments where the sacrificial substrate is an iron strip, the iron strip can be removed using a selective acid etch. The oxidization etch is tailored such that the iron strip is removed, but the materials of the electrode assembly are not affected.

These cochlear implant systems bypass the defective or missing hair cells located in the cochlea by presenting electrical stimulation directly to the ganglion cells in the cochlea. This electrical stimulation is supplied by an electrode array which is implanted in the cochlea. The ganglion cells then generate nerve impulses which are transmitted through the auditory nerve to the brain. This leads to the perception of sound in the brain and provides at least partial restoration of hearing function (50).

5.2.12 *Wear of the Total Intervertebral Disc Prosthesis*

Backache is a widespread disease which is mainly caused by degenerative changes of the intervertebral disc and the rise of sedentary activities (51).

Meanwhile, a rising number of intervertebral disc prostheses are implanted and therefore it is necessary to follow up on the wear behavior of total intervertebral disc prosthesis since wear particles can result in foreign body reactions and loosening of the implant from the corresponding bone.

The wear behavior of a tribological system with material combinations of the total intervertebral disc prosthesis InDisc (Institute of Medical and Polymer Engineering) was studied (51). This system consisted of two laser sintered titanium alloy plates with an elastomer in between them. A translatory oscillation test bench was used for the test, which simulates the kinematic and the physiological environment of the lumbar spine.

In the interface between the elastomer (Silopren LSR 2670) and the laser sintered titanium alloy (Ti_6Al_4V), abrasion cords out of silicone rubber with a maximal area of $500 \times 700\ \mu m^2$ were generated as well as individual particles with a size of 10×20 to $40 \times 50\ \mu m^2$. In the second prosthesis interface of Ti_6Al_4V and bone, bone particles with an area of 0.5×0.5 to $1 \times 1\ \mu m^2$ were detected. The wear particles and the amount of wear volume influence the biological response of a total intervertebral disc prosthesis in the body (51).

5.2.13 *Hand-Actuated Retention Catheter*

Catheters are tube-like devices which are inserted into a portion of a person's body in order to transport fluids, such as liquids, gases, and sometimes semisolids, in or out of that particular portion of the body (52). For instance, urinary catheters are used to transport urine collected in the bladder out of the body via the urinary tract. Other types of catheters, such as gastronomy devices, transport fluids into and out of various segments of the gastrointestinal system, primarily the stomach.

In order to provide a means of retaining the catheter within the body, inflatable bag catheters have been introduced (52). Also, an elongated catheter has been developed with a secondary lumen for inflating a retention balloon at a distal end of the catheter once the distal end is positioned within the body. Such catheters are now generally referred to as *Foley* catheters (53).

In spite of the many practical uses for these devices today, they do have limitations, a few of which are discussed below (52):

1. They are difficult for untrained individuals to use, especially relatively untrained nursing home attendants and/or patients who may wish to care for their own needs.
2. Because the fluid to expand the balloon is delivered from an external source, it is possible to burst the balloon by injecting too much fluid. It will be appreciated that this creates a safety concern.
3. It is possible that the secondary lumen, which communicates with the cavity of the expandable balloon, may become clogged during use of the catheter, thereby creating a problem when it becomes time to deflate the expandable balloon and remove the catheter.
4. The outer surface of the catheter leading to the balloon can irritate internal surfaces of the body with which it comes into contact, thereby creating inflamed areas which can be painful to the patient and may be more susceptible to bacterial infection.
5. The conduit portion of the catheter immediately adjacent to the expandable balloon does not fully engage or conform to the internal surfaces of the body approximate the expandable balloon. In the case of a urinary catheter, this often enables urinary fluids from the bladder to pass through the sphincter at the proximal end of the urinary passageway, thereby allowing urinary fluids to leak out of the bladder into the urinary passageway, thereby creating a risk of infection for the catheterized patient.
6. The cost of manufacturing traditional Foley catheters has been influenced by the need to use a significant amount of hand labor to make the devices.

Therefore, there is a need for a retention catheter, as well as methods for making and using the same, which will address these and other problems associated with prior art devices and methods.

A hand-actuated retention catheter has been developed (52). This device contains a tube having outer and inner surfaces, the inner surface defining an inner lumen, an overcoat layer encircling the tube, the overcoat layer having interior and exterior surfaces, a cavity interposed between the tube and overcoat layer, encircling the tube and being defined by portions of the outer surface of the tube and portions of the interior surface of the overcoat layer, and a fluid within the cavity.

The overcoat layer includes an expandable balloon or expandable balloon section and a squeeze bulb or fluid reservoir section interconnected by a constricting conduit or catheter sleeve section. The overcoat layer is joined to the outer surface of the tube at distal and proximal ends of the cavity. The cavity includes an expandable balloon portion and a fluid reservoir portion interconnected by a catheter sleeve portion. The catheter sleeve section of the overcoat layer defines a narrowing in the cavity through which fluid passing from the fluid reservoir portion to the expandable balloon portion thereof must pass and preferably includes restriction means for restricting the passage of fluid from the expandable balloon portion to the fluid reservoir portion via the catheter sleeve portion of the cavity, wherein a sufficient amount of the fluid can pass from the fluid reservoir portion, through the catheter sleeve portion, and into the expandable balloon portion when the fluid reservoir portion is compressed, so that the expandable balloon portion of the cavity is enlarged and the outer surface of the overcoat layer proximate the balloon section thereof is also enlarged. Although the catheter sleeve portion of the cavity, and the corresponding catheter sleeve section of the overcoat layer, can have virtually any practical length, for the commercial embodiments presently envisioned, they will preferably have lengths of at least about 2 *in*.

The hand-actuated retention catheter is fabricated by the following steps of (52,54):

1. Providing a silicone rubber tube,
2. Initially coating portions of an outer surface of such silicone rubber tube with a bond-preventing agent in a plurality of dipping steps, wherein the tube is immersed into the bond-preventing agent to a desired depth for a desired length of time, and subsequently removed, said plurality of

dipping steps being automated in series by mechanized automation means including computer control means, wherein the desired depth and the length of time for each of the plurality of dipping steps is prescribed so that a residual coating of bond-preventing agent remains on portions of the silicone rubber tube following said plurality of dipping steps, said residual coating having a variable thickness as a result of a variation between the depth of any two of said plurality of dipping steps, and

3. Subsequently coating the silicone rubber tube and the residual coating of bond-preventing agent with a polymeric bonding composition containing silicone rubber to form a shaped overcoat layer, wherein the shape of the overcoat layer results in part from the variable thickness of the residual coating.

5.2.14 *Medical Catheter*

Silicone rubber is physiologically inactive and has little response to tissue when it is touched by a living organism (55,56). Thus, silicone rubber has been utilized as a material for medical instruments such as various catheters for medical appliances.

A medical catheter is a tube which is inserted into a body cavity such as cavum thoracis or abdominal cavity, a lumen part such as alimentary canal or ureter, or a blood vessel, to be used for draining body fluid or injection infusion of drug solution, nutritional tonic, and contrast media. A medical catheter necessitates wound-resistance (tear-resistance), kink-resistance (tensile strength), transparency, and flexibility (stretching-extensibility), in addition to biocompatibility.

Some examples of the uses of medical catheters are, specifically, as a drainage tube of an evacuator for draining and removing postoperative sanguis or pus, and a tube for postoperative nutrition such as percutaneous endoscopic gastrostomy (PEG). In addition, in order to produce silicone rubber in an extra-fine tube shape for the catheter, silicone rubber composition as a silicone rubber material should have extrusion moldability.

As materials for a medical catheter, soft poly(vinyl chloride) (PVC) has been generally used in addition to silicone rubber. Although silicone rubber is superior to PVC, for biocompatibility and flexibility, it is required to improve the strength of silicone rubber such as tear strength or tensile strength, and in particular, tear strength (55).

If the tear strength is not sufficient, then the catheter may be broken by scarring caused by a needle or cutting tool when carrying out a surgical operation, or, if tensile strength is not sufficient, then the catheter may be bent to be yielded and kinked, and thereby distribution in the catheter of body fluid to be drained or drug solution to be infused may be stagnated.

Therefore, various methods have been proposed in the previous literature (57–60) in order to improve the tear strength and the tensile strength of silicone rubber.

A specific method to provide high tear characteristics to silicone rubber is the addition of an inorganic filler such as silica fine particles. Also, the crosslinking density in the silicone rubber system can be improved. The improvement of the tear characteristics by an uneven distribution of the crosslinking density is because the area of high-crosslinking density serves as resistance against the tear internal force.

The reason why tear strength is improved by a silica filler can be thought of as follows: That is, as a result of improvement of dispersibility of silica filler, the interface between silica filler and rubber matrix increases to increase the rubber molecular chain which is affected by a silica filler. Thus, the reinforcing effect by silica filler increases to improve mechanical strength. As for the rubber molecular chain which is affected by silica filler, molecular mobility decreases due to the interaction with silica filler, and as a result, it will be harder than the portion having high molecular mobility. In the behavior of tearing of silicone rubber, if tearing stress is added to a hard structure during initial cracks growth and spreads, then it serves as a resistance, and as a result, the tear strength increases.

With regard to curable silicone rubber composition, silicone rubber can be obtained by the following process: After heating the composition at a temperature ranging from 140°C to 180°C for a time period ranging from 5 *min* to 15 *min* (i.e., the primary curing), subjecting the resultant composition to a post-baking (secondary curing) at 200°C for 4 *h* (55).

In detail, the curable silicone rubber composition is pressed under a condition of a temperature of 170°C, 10 *MPa*, for 10 *min* into a sheet of 1 *mm* in thickness. The resultant sheet is subjected to a primary curing, subsequently the resultant sheet is heated at a temperature of 200°C for 4 *h* to perform a secondary curing, and the resultant silicone rubber sheet is shaped into crescent type test pieces according to JIS K6252 (2001) (61), therewith the tear strength of the crescent type test pieces according to JIS K6252 (2001) is measured.

The silicone rubber sheet is punched with a ring cutter having a center peripheral length of 50 *mm* and a width of 1 *mm*, into a silicone rubber ring.

The first vinyl group containing straight chain organopolyoxosilane was synthesized, in accordance with the chemical formula shown in Figure 5.5. In detail, the preparation runs as follows (55):

Preparation 5–1: Into a 300 *ml* separable flask being substituted by Ar gas, equipped with a condenser tube and a stirring wing, octamethylcyclotetrasiloxane (compound 1 in Figure 5.5) in an amount of 74.7 *g* (252 *mmol*), 2,4,6,8-tetramethyl 2,4,6,8-tetravinylcyclotetrasiloxane (compound 4 in Figure 5.5) in an amount of 0.086 *g* (0.25 *mmol*) and potassium siliconate in an amount of 0.1 *g* were added, and then it was heated and agitated at 120°C for 30 *min*. An increase of the viscosity could be confirmed.

Thereafter, it was heated to 155°C, and stirring was continued for 3 *h*; 3 *h* later, 1,3-divinyltetramethyldisiloxane in an amount of 0.1 *g* (0.6 *mmol*) was added, it was further stirred at 155°C for 4 *h*.

Then, after the reaction mixture was diluted with 250 *ml* of toluene, the resultant mixture was washed with water three times. The organic layer after washing was washed with 1.5 *l* of methanol to be reprecipitated and purified, thereafter the oligomer and polymer were separated therefrom. The resultant polymer was depressurized at 60°C overnight to be dried, and thereby the second vinyl group containing straight chain organopolyoxosilane was obtained ($M_n = 277,734, M_w = 573,906$, IV value 0.89 $dl\ g^{-1}$).

Furthermore, the preparation of the curable silicone rubber composition can be done as follows (55):

Preparation 5–2: Into 100 parts by weight of the first vinyl group containing straight chain organopolysiloxane, 70 weight parts of the silica filler was added, then the resultant mixture was kneaded to prepare a masterbatch.

Subsequently, 0.5 weight parts of platinum was added thereto and kneaded until the mixture became homogeneous, thereafter 2.0 wait parts

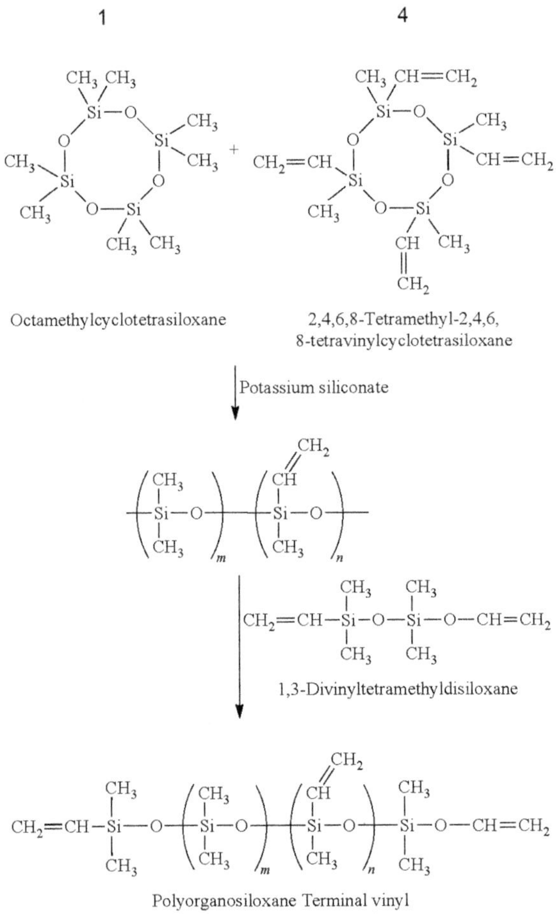

Figure 5.5 Preparation of a curable silicone rubber composition (55).

of the straight chain organohydrogenpolysiloxane was added thereto and kneaded to prepare a curable silicone rubber composition.

In summary, the silicone rubber obtained by curing the curable silicone rubber composition excels in tear strength. Accordingly, each of the compact made of the curable silicone rubber composition and the medical tube constituted from the compact excels in mechanical strength such as tear strength. So, it is possible to provide a medical catheter made of silicone rubber having excellent scar resistance (55,56).

5.2.15 *Silicone-Coated Stents*

Nitinol (NiTi) is an alloy of nickel and titanium which has been used in the construction of medical devices, including self-expanding stents to brace the inside circumference of a tubular passage such as cardiac, peripheral vascular, biliary, and gastrointestinal (62).

NiTi strips modified by the plasma etching were subjected to a liquid silicone rubber coating after plasma treatment or primer treatment and then thermally treated at 150°C for 24 *h* to evaporate the solvent.

The interfacial bonding of these silicone-coated stents was characterized by field emission SEM and anodic potentiodynamic polarization tests. The coaxial in-lens secondary electron imaging revealed a partial bonding between silicon rubber and nitinol alloy strips both in the cases of primer-treatment and plasma-treatment. Several strands between the polymer and stent substrate were observed in both cases and it seems that the bonding strength is the same.

However, potentiodynamic polarization tests showed that the corrosion rates of the plasma-treated nitinol stents were less than half of that of the primer-treated stents, indicating a higher corrosion resistance. The enhanced corrosion resistance may be attributed to the modification of the physical properties of the surface of the stents by plasma treatments, making the oxide layer more effective than that of the primer-treated stents (62).

5.2.16 *Suture Sleeve*

Implantable medical electrical stimulation and/or sensing leads are well known in the fields of cardiac stimulation and monitoring, including cardiac pacing and cardioversion/defibrillation, and in other fields of electrical stimulation or monitoring of electrical signals or other physiologic parameters (63).

In the field of cardiac stimulation and monitoring, endocardial leads are placed through a transvenous route to locate one or more stimulation and/or sense electrode along or at the distal end of the lead body in a desired location within a chamber of the heart or within a blood vessel of the heart. Epicardial leads are routed from a subcutaneous site to the epicardium of the heart to dispose one or more stimulation and/or sense electrode along or at the distal end of the lead body against the epicardial surface or into the myocardium. A pacemaker implantable pulse generator or implantable cardioverter/defibrillator implantable pulse generator or monitor is coupled to the heart through one or more of such endocardial or epicardial leads. For convenience, epicardial and endocardial leads for monitoring, pacing or cardioversion/defibrillation are collectively referred to hereafter as cardiac leads unless they are explicitly identified.

A suture sleeve is shown in Figure 5.6. A method of fabrication

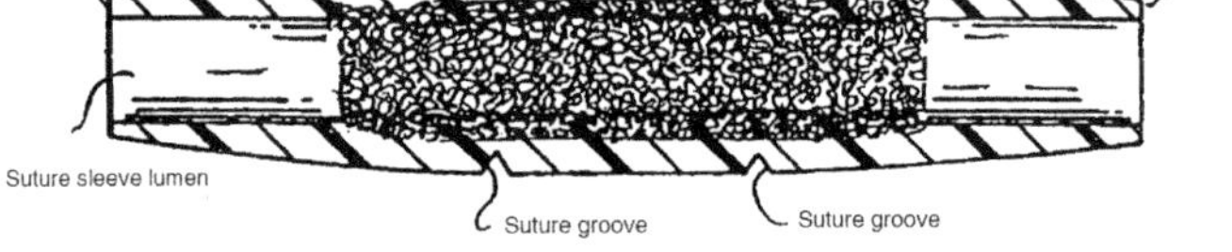

Figure 5.6 Suture Sleeve (63).

of a suture sleeve is as follows (63):

1. Provide a mandrel with a mandrel diameter correlated to lead body diameter,
2. Mix particulate media and suspension agent into a slurry coat mandrel with release agent,
3. If necessary, apply slurry over an insert length of the coated mandrel and cure to form a sleeve insert,

4. Mold silicone rubber sleeve body over sleeve insert and mandrel,
5. Remove mandrel and sleeve body from mold, optionally form a slit, and
6. Remove mandrel from sleeve lumen.
7. Optionally apply adhesive to sleeve lumen.

The suspension agent can be one of a solvent thinned liquid silicone adhesive or a two-part liquid silicone rubber that sets up, when mixed together, at room temperature, or a solvent-based urethane adhesive that bonds to or can be bonded to silicone rubber (63).

5.2.17 *Silicone Tubings*

Silicone articles, such as tubing, may be used in a variety of industries such as the medical industry, pharmaceutical industry, and biopharmaceutical industry (64). Gamma radiation may be used for the sterilization of products for biopharmaceutical applications. However, certain issues may arise using conventional silicone-based materials for biopharmaceutical applications.

It was discovered that after sterilization with at least one dose of irradiation, acidic species are formed, which could be potentially harmful when a fluid is in contact with the silicone-based composition. For instance, after gamma irradiation, the pH of the silicone-based composition after exposure to water is less than 5 for a silicone-based composition including a silicone polymer without an acid neutralizer. However, the inclusion of at least one acid neutralizer with the silicone polymer would increase the pH of the silicone-based composition to at least about 5, such as at least about 6 after at least one dose of irradiation, when measured by a pH meter.

A silicone-based composition has been developed that contains a silicone polymer and at least one acid neutralizer, wherein the silicone-based composition has a pH of at least about 5 after at least one dose of irradiation, when measured by a pH meter (64).

The pH of various compositions after γ-irradiation is shown in Table 5.2.

The pH of compositions with calcium oxide after γ-irradiation is shown in Table 5.3.

Table 5.2 pH after γ-irradiation (64).

Composition	pH
LSR without filler	4.06
LSR with silica filler	3.74
LSR with 1% CaO filler	10.03

Table 5.3 pH after γ-irradiation (64).

CaO/[%]	pH	CaO/[%]	pH
1.00	10.68	0.50	10.61
0.30	10.06	0.10	9.40

5.2.18 *Fresnel Lenses*

Fresnel lenses have apertures and focal lengths comparable to conventional lenses (65). Because Fresnel lenses are typically thinner and lighter than conventional lenses of similar performance features, e.g., aperture and/or focal length, replacing conventional lenses in head-mounted displays with Fresnel lenses can reduce the size and weight of the head-mounted displays. However, Fresnel lenses suffer from optical artifacts associated with Fresnel pattern, and thus, the use of Fresnel lenses in imaging applications is limited.

LSR parts are used in aerospace application due to their stable properties (66). In order to compress the time to market a new product, it is necessary to develop an injection mold for LSR swiftly and effectively.

A rapid tooling technology can reduce the time to market compared to the conventional machining methods. An aluminum-filled epoxy resin mold is capable of replacing the conventional steel mold for small batch production using injection molding.

Here, an injection mold with heating element was developed using rapid tooling technology for LSR injection molding. It was found that the replication rate for the average microgroove depth and width of the aluminum-filled epoxy resin mold is about 90.5% and 98.9%, respectively. The transcription rate for the average microgroove depth and width of the LSR molded parts is about 91.5% and 99.2%, respectively.

The variations in the dimension of microgroove depth and the width of the LSR molded parts can be controlled within $\pm 1 \mu m$. The average surface roughness of the Al-filled epoxy resin mold was only increased by about 12.5 nm after 200 test runs of LSR injection molding (66).

A lens assembly and a method of making the assembly have been described (65). The lens assembly includes a first lens and a second lens slidably coupled with the first lens. The second lens includes a silicone material and has a Fresnel pattern surface. Also described is a display device including the lens assembly and an array of light-emitting devices coupled with the lens assembly for outputting light through the lens assembly (65).

5.3 Biomaterials

Biomaterials are materials which are used to replace a part or a function of the body in a safe, reliable, economical and physiologically acceptable manner (67). Polymers have been used as biomaterial for a long time due to their light weight and biocompatibility. Natural and synthetic rubbers are mainly used because of their inertness and bicompatible nature. Silicone rubbers are widely used and natural rubber as of yet has found only limited application.

The roles of silicone rubber and natural rubber as biomaterials have been detailed and reviewed (68,69).

Furthermore, both rubbers were described and their applications have been discussed in a monograph (67). Also, medical applications of silicones have been described in a monograph (70).

5.3.1 *Bioactive Peptides Grafted Silicone Dressings*

The need for bioactive dressings increases with the population aging and the prevalence of chronic diseases. In contrast, there are only a few dressings on the market which are designed to display a chosen bioactivity.

A simple method for silicone dressing functionalization with peptides has been presented (71). The grafting proceeds in a single step, by the incubation with a hybrid silylated peptide.

The surface-functionalization of silicone wound dressing with bioactive peptides was investigated (71). One of the challenges was to avoid multistep grafting reactions involving catalysts, solvents or toxic reagents, which are not suitable for the fabrication of medical devices at an industrial scale.

A covalent bonding is necessary to avoid the loss of the biological effect by progressive removal of the peptide in biological fluids generated by the wound. To solve these limitations, a strategy was developed that allows an easy and direct functionalization of silicone. This strategy relied on hybrid silylated bioactive peptides, which can chemoselectively react with plasma-activated silicone surfaces. Three hybrid peptides were synthesized with wound healing properties, which were grafted onto the commercially available silicone dressings Cerederm® and Mepitel®. The grafted dressings were evaluated *in-vitro*. They enabled a quicker scare recovery and extracellular matrix deposition with human dermal fibroblasts. These results were confirmed by *in-vivo* studies that were showing an enhanced wound healing on a pig skin (71).

5.3.2 *Antibacterial and Antibiofouling Clay Nanotube-Silicone Composites*

The prevention of bacterial adherence and biofilm development continues to be a major healthcare challenge (72). So, there is a need to improve the antimicrobial properties of medical devices. The production of hydrophilic silicone coatings with antibacterial and antibiofouling properties has great potential for use in contact lenses, dialysis membranes, bladder and central venous catheters,

Poly(dimethyl siloxane) (PDMS) was doped with halloysite nanotubes, and the PDMS halloysite nanotube composite surfaces were coated with PDMS poly(ethylene oxide) (PEO) and antibacterial agents. Halloysite nanotubes are naturally occurring clay nanotubes mined from abundant mineral deposits, making it an easily accessible nanomaterial. A halloysite nanotube is structured as a two-layered aluminosilicate and has a predominantly hollow nanotubular structure in the submicron range.

The composite material properties were examined using SEM, energy dispersive spectroscopy, water contact angle measurements, tensile testing, UV-Vis spectroscopy, and thermogravimetric analysis. The antibacterial potential of the PDMS-halloysite nanotube composites was compared to commercial urinary catheters using cultures of *E. coli* and *S. aureus*. Fibrinogen adsorption studies were also performed on the PDMS halloysite nanotube PEO composites.

The addition of halloysite nanotubes increased drug load during solvent swelling without reducing material strength. The hydrophilic properties provided by PEO were maintained after halloysite nanotube addition, and the composites displayed protein-repelling properties. Additionally, composites were superior over commercial catheters at inhibiting bacterial growth.

PDMS halloysite nanotube composites showed superiority regarding their efficacy at inhibiting bacterial growth in comparison to commercial antibacterial catheters. The research suggested that PDMS halloysite nanotube composites have potential as a coating material for antibacterial invasive devices and in the prevention of institutional-acquired infections (72).

5.3.3 *Biofunctionalization with Microgroove-Patterned Surface*

Silicone rubber implants have been widely used to repair soft tissue defects and deformities. However, poor biocompatibility can elicit capsule formation, usually resulting in prosthesis contracture and displacement in long-term usage (73).

A possible solution is represented by the modification of the silicone rubber surface to minimize hydrophobic interaction and to improve the cell adhesion. A variety of surface modification methods, such as coating with carbon nanotubes, plasma spraying, sintering, and electrochemical deposition, can reduce the surface hydrophobicity of silicone rubber, increase the adhesion and proliferation of fibroblasts, and significantly improve cytocompatibility (74–77)

Remodeling of a silicone rubber surface by C ion implantation could effectively improve cytocompatibility. This improvement was attributed to changes in surface characteristics, including surface chemistry, surface roughness, and wettability (78,79)

Then, the properties of silicone rubber materials with or without a microgroove-patterned surface and with or without carbon ion implantation were investigated (73).

Atomic force microscopy, X-ray photoelectron spectroscopy, and a water contact angle test were used to characterize the surface morphology and the physicochemical properties. The cytocompatibility was investigated by a cell adhesion experiment, immunofluorescence staining, a cell counting kit-8 assay, and SEM *in vitro*. The histocompatibility was evaluated by studying the inflammatory response and fiber capsule formation that developed after subcutaneous implantation in rats for 7 *d*, 15 *d*, and 30 *d in vivo* (73).

Parallel microgrooves were found on the surfaces of patterned silicone rubber and patterned C ion implanted silicone rubber. Irregular larger peaks and deeper valleys were present on the surface of silicone rubber implanted with C ions. The silicone rubber surfaces with microgroove patterns had stable physical and chemical properties and exhibited moderate hydrophobicity.

Patterned C ion implanted silicone rubber moderately increased the dermal fibroblast cell adhesion and growth, and its surface microstructure promoted orderly cell growth.

Furthermore, histocompatibility experiments on animals showed that both the anti-inflammatory and antifibrosis properties of patterned C ion implanted silicone rubber were slightly better than those of the other materials, and there was also a lower capsular contracture rate and less collagen deposition around implants made from patterned C ion implanted silicone rubber (73).

5.3.4 *Bionic Composites*

A microorganism metabolic activity can facilitate the formation of cellular material systems that have unusual mechanical and physical properties (80). In the living world, microorganisms are commonly used for preparing porous food by fermentation.

Here, carbon nanotubes, graphene nanoplatelets, and a mix of them are dispersed in liquid silicone rubber with single-cell fungi of commercial beer yeast. The fermentation of such microorganisms during the gelling of the silicone matrix results in bionic composites with buckled/collapsed cells that infer, as rationalized with an analytical model and excluded in an abiotic experimental comparison, auxetic properties.

The fermentation-assisted method that has been proposed for the realization of bionic composites involves a prereaction state where the solution containing yeast and sugar releases CO_2 bubbles that are trapped by the crosslinking agent in silicon rubber (SR), when the viscosity of the solution starts to increase with the polymerization degree (80).

Once the gelation starts, the accumulation of CO_2 gas, that needs to escape, generates pores and, as a result, porous SR and SR composite have been obtained.

An attempt has been made to reproduce *Saccharomyces cerevisiae* through a process called *budding*, where a daughter cell is initiated as growth from the mother cell.

The yeast cells can be completely removed by post-thermal annealing at 250°C of the bionic composite after their cure. From field emission scanning electron microscopy analysis it is possible to observe that the cells in the SR/CNTs composite are round and rather symmetrical, similar to conventional cell structure previously reported for SR foams. In contrast, the cells of the bionic SR/graphene nanoplatelets and SR/CNTs/graphene nanoplatelet composites are collapsed or buckled. The values of the pore size and the pore shape as well as the statistics from the analysis of the cross sections of the bionic samples have been detailed (80).

During stretching it was found that the Poisson's ratio of such composites changes sign, from negative to positive, and the variation of the electrical resistance is negative. In addition to the conductivity increment, a general increment of the stretchability and damage resistance with respect to the composites prepared by abiotic process is observed. Bionic composites, even if in their infancy, can thus be multifunctional and superior to their traditional/abiotic counterparts (80).

5.4 Pharmaceutical Compositions

Silicones are hydrophobic substances that offer unique cosmetic properties (81). They are tasteless, essentially odorless, non-greasy and non-stinging. They are often used as a base fluid in many personal care products, with excellent spreading and easy rubout and lubrication properties.

Volatile silicone compounds are a specific class of silicones used in dermal formulations to condition the skin, while reducing the greasy feel of other oils in the formulation. However, one of the principal drawbacks to the use of silicones in foam formulations is their known antifoaming effect, especially when included in the formulation in substantial quantities. Also, silicones are known for being efficient foam control agents and can prevent foam formation or cause foam to collapse rapidly.

Silicone fluids can, for example, enter into the foam lamella and displace the foam-stabilizing surfactants from the interphase. The foam lamellas are therefore destabilized and burst resulting in foam collapse. Thus, silicones are essentially contraindicated for the preparation of foamable carriers and compositions. Nevertheless, because of the favorable properties of silicones, there is still an unmet need to develop foamable formulations incorporating silicone that can produce easy to use good quality foam, especially in substantial quantities.

Pharmaceutical and cosmetic foams can be generated by manufacturing a suitable foamable carrier composition and loading the carrier in a pressurized valved canister with an appropriate propellant. Upon expelling the canister contents, a foam can be released. The type, nature and quality of the foam depends, inter alia, on the carrier composition, the active agent, the propellant and the method of manufacture and storage.

Compositions formulated using a base containing an oil combined with liquid silicones to produce waterless formulations have been investigated and developed as pharmaceutical and cosmetic waterless carriers suitable for delivery of a wide range of active agents despite the defoaming properties of silicones (81). In particular, such carriers that are substantially a single phase and, in some embodiments, are substantially free of particles.

Moreover, such carriers are ideal for oil soluble active agents and can nevertheless carry as a homogenous suspension substantial amounts of oil insoluble active agents.

Silicones such as dimethicone, simethicone and simethicone emulsion have found wide application in the manufacture of pharmaceuticals due to their efficient antifoam properties (81). Dimethicones and simethicones are used as antifoams in anti-flatulent or anti-acid formulations.

Silicones in these products help to suppress the formation of foam in the stomach without modifying the gastric pH. This is not surprising, as silicones, with their low surface tension, in particular when compounded with silicon dioxide, are known to destroy foams in many applications, e.g., in petrol, paper pulp or food processing. This makes them contraindicated for use in foamable carriers and compositions and especially in waterless or substantially waterless carriers, compositions, and foams.

With regard to shape and size, silicones that do not inhibit the hydrophobic-hydrophobic interactions of the oil are thought to produce more stable foams. Exemplary silicones of appropriate size and shape include cyclic silicones.

Exemplary cyclic silicones include, without limitation, cyclomethicone, cyclotetrasiloxane, cyclopentasiloxane, and cyclohexasiloxane. These compounds are shown in Figure 5.7.

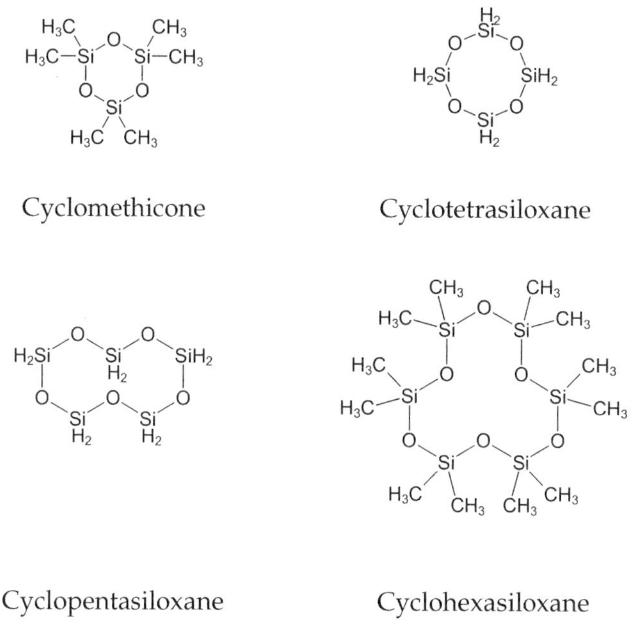

Figure 5.7 Cyclic silicones.

Other examples of silicones of appropriate size and shape are silicones having side groups or side chains. Examples of silicones having branched side chains are stearyl dimethicone and phenyl trimethicone, cetyl dimethicone, and caprylyl methicone. These compounds are shown in Figure 5.8.

A waterless foamable pharmaceutical composition has been presented. The composition includes a foamable carrier and at least

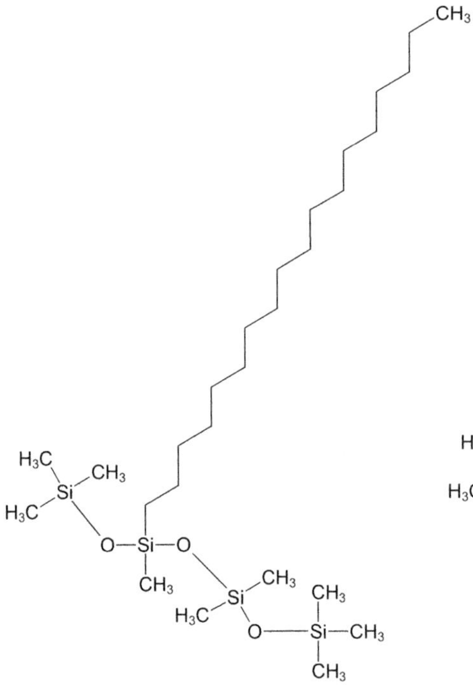

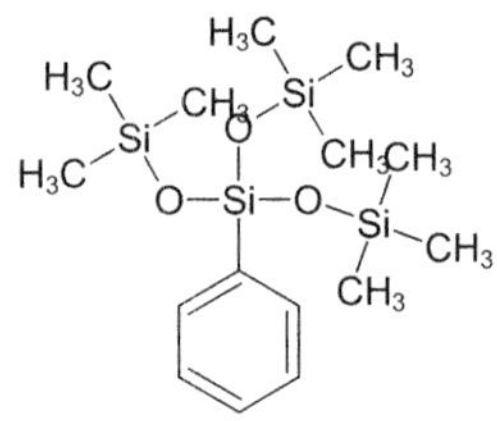

Stearyl dimethicone Phenyl trimethicone

Figure 5.8 Silicones with branched side chains.

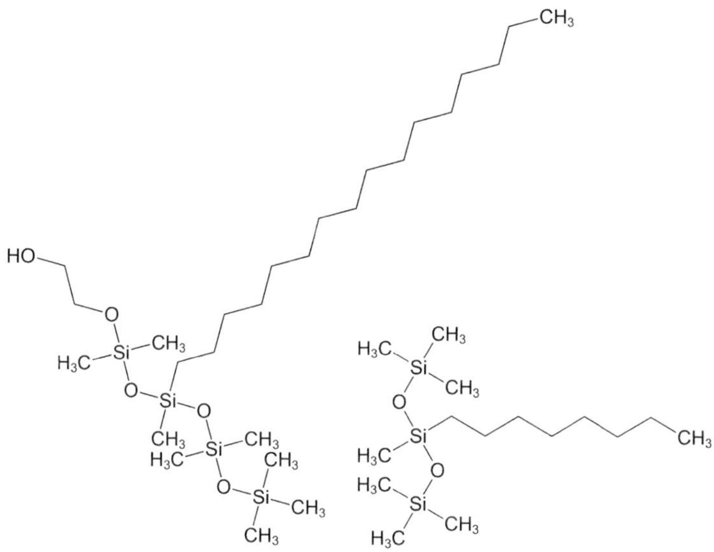

Cetyl dimethicone Caprylyl methicone

Figure 5.8 (cont.) Silicones with branched side chains.

one liquefied or compressed gas propellant. The foamable carrier includes (81):

1. About 60% to 95% of at least one liquid oil,
2. A silicone, and
3. About 0.01% to about 25% by weight of a monoglyceride, diglyceride, or triglyceride, or a mixture thereof, wherein the side chain of the monoglyceride, diglyceride, or triglyceride is a saturated hydrocarbon.

The foamable carrier is substantially a single phase. The ratio of the foamable carrier to the propellant is from 100:03 to 100:35.

The foamable carriers may further contain an emollient. Examples of emollients for use in the compositions include, cocglycerides, PPG-15 stearyl alcohol, octyl dodecanol, isopropyl myristate, diisopropyl adipate, cetyl ethylhexanoate, isohexadecanol and mixtures thereof. Some emollients are shown in Figure 5.9.

A method for preparing a substantially single phase, waterless foamable pharmaceutical composition includes the steps of (81):

1. Combining about 60% to about 95% by weight of at least one liquid oil with at least one foam-stabilizing agent at a temperature of at least about 50°C,
2. Rapidly cooling the combination of one liquid oil and at least one foam-stabilizing agent to less than 40°C,
3. Adding a silicone to the cooled combination to make a prefoam formulation, and
4. Combining the prefoam formulation with a compressed gas propellant in a weight ratio of 100:3 to 100:35.

In this method, the foam-stabilizing agent is selected from the group consisting of about 0.01% to about 25% of at least one surface-active agent alone or in combination with a foam adjuvant; about 0% to about 5% of at least one polymeric agent alone or in combination with a foam adjuvant and mixtures thereof.

Some compositions may contain at least one hydrophobic solvent. Examples of hydrophobic solvents are shown in Table 5.4.

Preferably, the pharmaceutical composition has the following property: a foam quality of at least good up to excellent; and at least one other property selected from: Specific gravity in the range

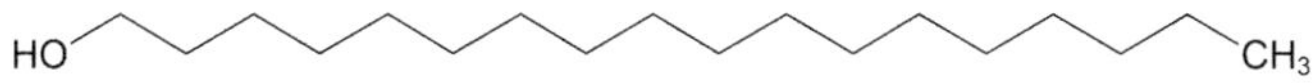

Stearyl alcohol

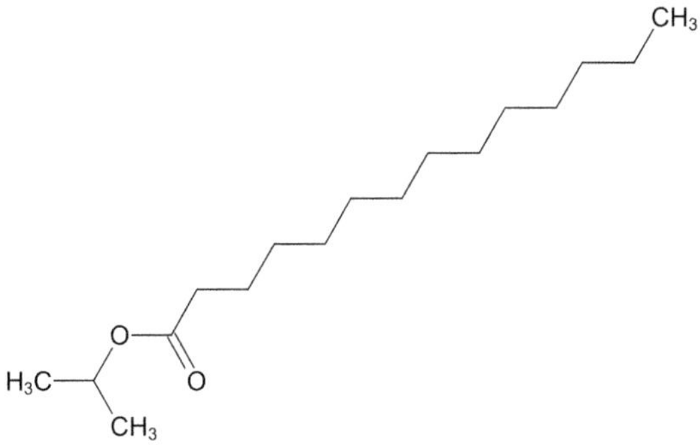

Isopropyl myristate

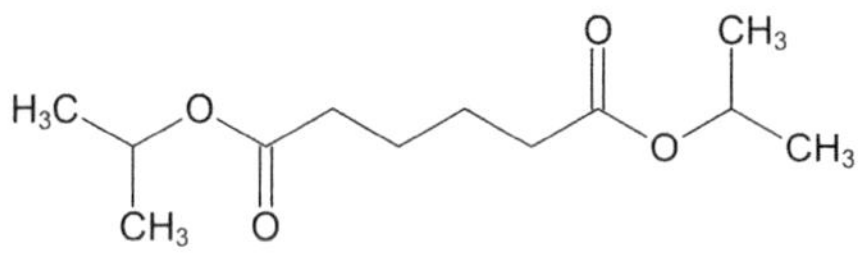

Diisopropyl adipate

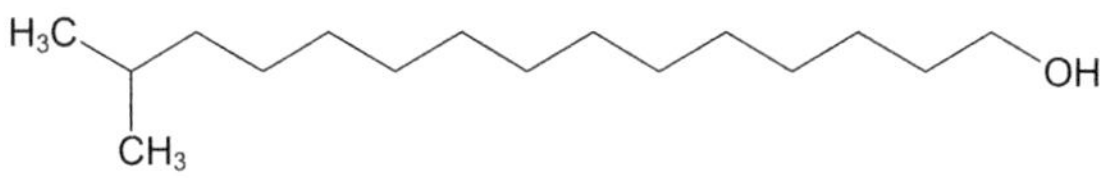

Isohexadecanol

Figure 5.9 Emollients.

Table 5.4 Hydrophobic solvents (81).

Compound	Compound
Mineral oil	Isopropyl palmitate
Isopropyl isostearate	Diisopropyl adipate
Diisopropyl dimerate	Maleated soybean oil
Octyl palmitate	Cetyl lactate
Cetyl ricinoleate	Tocopheryl acetate
Acetylated lanolin alcohol	Cetyl acetate
Phenyl trimethicone	Glyceryl oleate
Tocopheryl linoleate	Wheat germ glycerides
Arachidyl propionate	Myristyl lactate
Decyl oleate	Ricinoleate
Isopropyl lanolate	Pentaerythrityl tetrastearate
Neopentyl glycol dicaprylate	Isononyl isononanoate
Isotridecyl isononanoate	Myristyl myristate
Triisocetyl citrate	Octyl dodecanol
Olive oil	Corn oil
Soybean oil	Canola oil
Cottonseed oil	Coconut oil
Sesame oil	Sunflower oil
Borage seed oil	*Syzigium aromaticum* oil
Hempseed oil	Herring oil
Cod liver oil	Salmon oil
Flaxseed oil	Wheat germ oil

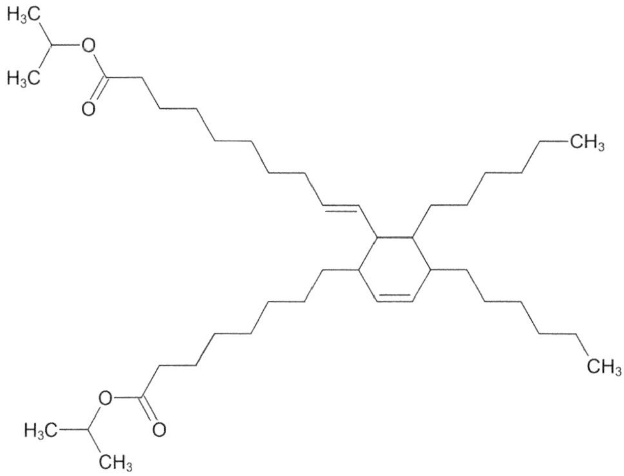

Diisopropyl dimerate

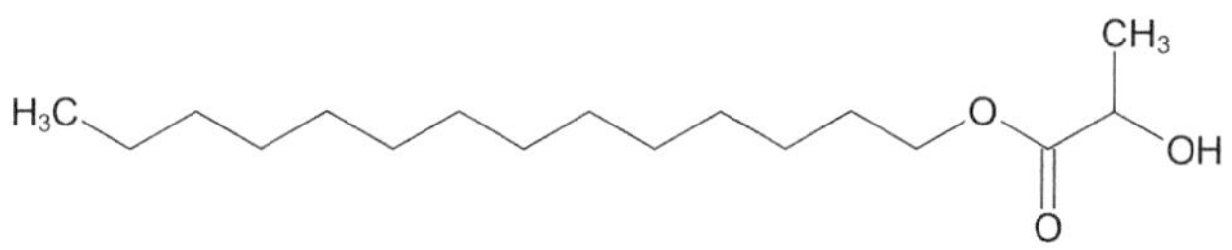

Myristyl lactate

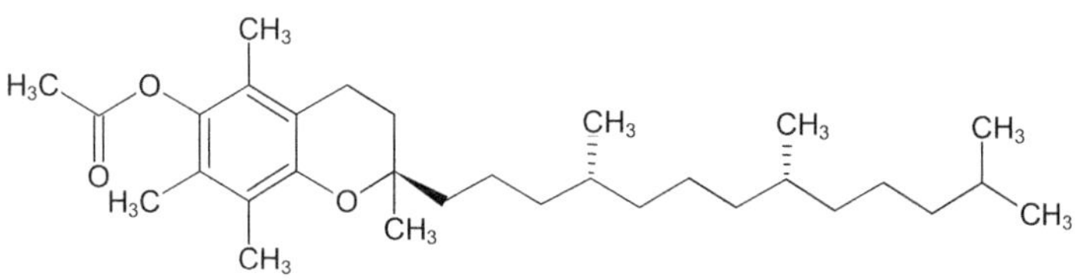

Tocopheryl acetate

Figure 5.10 Hydrophobic solvents.

of about 0.05 $g\,ml^{-1}$ to about 0.20 $g\,ml^{-1}$, a foam texture of a very fine creamy foam consistency to a fine bubble structure consistency; a sustainability of more than 95% for at least 1 *min* upon release thereof to a surface from an aerosol can capable of withstanding at least one freeze thaw cycle without substantial change, having a mean bubble size of less than about 200 μm, and a compatibility with at least one active agent.

The method for preparing a substantially single phase, waterless foamable pharmaceutical composition is useful for preparing the pharmaceutical compositions and foamable carriers (81).

References

1. W.R. Eichhorn and K.G. Wyrobek, Electrically conductive nanotube composite sensor for medical application, US Patent Application 20 150 044 656, assigned to 7-Sigma, Inc., Minneapolis (MN), February 12, 2015.
2. Z.-M. Dang, M.-J. Jiang, D. Xie, S.-H. Yao, L.-Q. Zhang, and J. Bai, *Journal of Applied Physics*, Vol. 104, p. 024114, 2008.
3. Wikimedia Commons, File:shore a vs asker c scales.png — wikimedia commons, the free media repository, 2014. [Online; accessed 28-January-2019].
4. W.R. Eichhorn, R. Duda, and K.G. Wyrobek, Pressure sensor with a deformable electrically resistive membrane, US Patent 9 534 972, assigned to 7-Sigma Inc. (Minneapolis, MN), January 3, 2017.
5. W.R. Eichhorn, R.M. Duda, and K.G. Wyrobek, Responsive tool with sensors, US Patent Application 20 150 064 675, assigned to 7-Sigma, Inc., Minneapolis (MN), March 5, 2015.
6. G.-Z. Chen, I.-S. Chan, L.K. Leung, and D.C. Lam, *Medical Engineering & Physics*, Vol. 36, p. 1134, 2014.
7. S. Higaki, T. Ohtake, and M. Yokoyama, Rod template, US Patent 10 080 593, assigned to Nishijima Medical Co., Ltd., September 25, 2018.
8. B. Yuan, Method and device for making silicone rubber cupping appliance, US Patent 9 573 305, assigned to Green Island International TCM Group Limited (Hong Kong, HK), February 21, 2017.
9. R. Xu, Y. Bai, J. Zhao, H. Xia, Y. Kong, Z. Yao, R. Yan, X. Zhang, X. Hu, M. Liu, Q. Yang, G. Luo, and J. Wu, *Journal of Tissue Engineering and Regenerative Medicine*, Vol. 12, p. e905, 2017.
10. Wikipedia contributors, Wnt signaling pathway — Wikipedia, the free encyclopedia, https://en.wikipedia.org/w/index.php?title=Wnt_

signaling_pathway&oldid=886738880, 2019. [Online; accessed 12-March-2019].

11. P. Li, S. Jiang, Y. Yu, J. Yang, and Z. Yang, *Journal of the Mechanical Behavior of Biomedical Materials*, Vol. 49, p. 220, 2015.
12. Wikipedia contributors, Cleidocranial dysostosis — Wikipedia, the free encyclopedia, https://en.wikipedia.org/w/index.php?title=Special:CiteThisPage&page=Cleidocranial_dysostosis&id=883795832, 2019. [Online; accessed 11-March-2019].
13. R. Horch, D.W. Schubert, S. Werner, M. Hedenqvist, and R.T. Olsson, Silicone breast implant with reinforcing fibers, WO Patent 2 017 093 181, assigned to Friedrich-Alexander-Universität Erlangen-Nürnberg, Universitätsklinikum Erlangen, June 08, 2017.
14. A. Colas and J. Curtis, Silicone biomaterials: History and chemistry in B.D. Ratner, ed., *Biomaterials Science: An Introduction to Materials in Medicine*, chapter 2.3, pp. 80–85. Elsevier Academic Press, Amsterdam Boston, 2nd edition, 2004. Part I, Materials Science and Engineering.
15. A.U. Daniels, *Swiss Medical Weekly*, Vol. 142, p. 1, 2012.
16. D.W. Schubert, M. Lämmlein, and H. von Hanstein, *Polymer Testing*, Vol. 66, p. 292, 2018.
17. G. Ginsbach, L.C. Busch, and W. Kühnel, *Plastic and Reconstructive Surgery*, Vol. 64, p. 456, 1979.
18. A.B. Birkefeld, R. Bertermann, H. Eckert, and B. Pfleiderer, *Biomaterials*, Vol. 24, p. 35, 2003.
19. H.J. Brandon, K.L. Jerina, C.J. Wolf, and V. Young, *Aesthetic Surgery Journal*, Vol. 19, p. 381, 1999.
20. H.J. Brandon, K.L. Jerina, C.J. Wolf, and V. Young, *Aesthetic Surgery Journal*, Vol. 20, p. 122, 2000.
21. L. Pinchuk, J.B. Martin, Jr., and A.A. Maurin, Breast prostheses, US Patent 5 376 117, assigned to Corvita Corporation (Miami, FL), December 27, 1994.
22. V. Loria, Hair implants comprising enhanced anchoring and medical safety features, US Patent 9 993 334, assigned to Lorstan Pharmaceutical, LLC (Doral, FL), June 12, 2018.
23. S.A. Kaplan, D.E. Chung, R.K. Lee, S. Scofield, and A.E. Te, *International Journal of Clinical Practice*, Vol. 66, p. 1052, 2012.
24. P.R. Chellini, R. Pirmez, P. Raso, and C.T. Sodré, *International Journal of Trichology*, Vol. 7, p. 182, 2015.
25. J.M. Hirshburg, P.A. Kelsey, C.A. Therrien, A.C. Gavino, and J.S. Reichenberg, *The Journal of Clinical and Aesthetic Dermatology*, Vol. 9, p. 56, July 2016.
26. I.G. Motofei, D.L. Rowland, M. Manea, S.R. Georgescu, I. Păunică, and I. Sinescu, *Clinical Drug Investigation*, Vol. 37, p. 511, June 2017.
27. S.A. Habibullah Aktas, E.B. Türkoglu, and Ö. Sevik, *Journal of Clinical and Diagnostic Research: JCDR*, Vol. 10, p. WD01, 2016.

28. N.N., Banned devices, Code of Federal Regulations 895, U.S. Food and Drug Administration, Silver Spring, MD, 2018.
29. M.A. Erlich and A. Parhiscar, *Facial Plastic Surgery*, Vol. 19, p. 325, 2003.
30. N. Fanous and A. Yoskovitch, *Canadian Journal of Plastic Surgery*, Vol. 11, p. 161, 2003.
31. I. Niechajev, *Aesthetic Plastic Surgery*, Vol. 36, p. 917, August 2012.
32. A. Vanhoestenberghe and N. Donaldson, *Journal of Neural Engineering*, Vol. 10, p. 031002, 2013.
33. J.P. Chavoin, B. Chaput, I. Garrido, B. Moreno, M. Dahan, and J.L. Grolleau, *Annales de Chirurgie Plastique et Esthetique*, Vol. 61, p. 694, October 2016.
34. L. Nayyer, G. Jell, A. Esmaeili, M. Birchall, and A.M. Seifalian, *Advanced Healthcare Materials*, Vol. 5, p. 1203, 2016.
35. A.M.D. Wan, T.A. Moore, and E.W.K. Young, *JoVE (Journal of Visualized Experiments)*, Vol. N.A., p. e55175, 2017.
36. A. Inamdar, N. Adhe, S. Shende, R. Uddanwadiker, and S. Lulay, *International Journal of Pharma Medicine and Biological Sciences*, Vol. 5, p. 81, 2016.
37. P.F. Zheng, Q.S. You, Q. Li, H.Y. Deng, I.Y.H. Wong, and X.Y. Peng, *PLOS One*, Vol. 13, p. 1, March 2018.
38. Wikipedia contributors, Eosinophil — Wikipedia, the free encyclopedia, https://en.wikipedia.org/w/index.php?title=Eosinophil&oldid=883890127, 2019. [Online; accessed 11-March-2019].
39. M. Hogan, I.-Y. Yang, and Y. Sun, Composition and method of DNA marking elastomeric material, US Patent Application 20 180 105 873, assigned to APDN (B.V.I.) Inc., April 19, 2018.
40. G.M. Church, Y. Gao, and S. Kosuri, *Science*, Vol. 334, 2012.
41. F.C. Bancroft and C. Clelland, DNA-based steganography, US Patent 6 312 911, assigned to Icahn School of Medicine at Mount Sinai, November 6, 2001.
42. J. Stieghorst and T. Doll, *Additive Manufacturing*, Vol. 24, p. 217 , 2018.
43. L. Rubin, *Biomaterials in Reconstructive Surgery*, Mosby, St. Louis, 1983.
44. P. Hron, *Polymer International*, Vol. 52, p. 1531, 2003.
45. NuSil Tech., Product profile MED-4735, Product description, NuSil Technology LLC, Carpenteria, CA, 2004.
46. NuSil Tech., Product profile MED-4750, Product description, NuSil Technology LLC, Carpenteria, CA, 2018.
47. K.J.D.A. Buijssen, J.J.H. Oosterhof, L. Basil, M. Waters, M.A. Duits, H.J. Busscher, H.C. van der Mei, and B.F.A.M. van der Laan, *Clinical Otolaryngology*, Vol. 42, p. 1235, 2017.

48. J.J.H. Oosterhof, L. Basil, M. Waters, B.F.A.M. van de Laan, and H.J. Busscher, Biofilm formation on voice prostheses with decreased surface roughness in J.J.H. Oosterhof, ed., *Strategies to Decrease Biofilm Formation on Voice Prostheses*, chapter 6, pp. 83–93. University of Groningen, 2008.
49. L. Lim, D.N. Arnholt, J.T. Eggert, A. de Kock, P.E. Zarembo, and C.A. Hartemink, Implantable medical leads having oscillating cable conductor lumens, US Patent Application 20 160 067 475, assigned to Cardiac Pacemakers Inc., March 10, 2016.
50. C. Thenuwara and T. Beerling, Cochlear electrode array, US Patent 9 056 196, assigned to Advanced Bionics, LLC (Valencia, CA), June 16, 2015.
51. C. Micheler, E. Krampe, P. Foehr, M. Haerst, and R. Burgkart, *Current Directions in Biomedical Engineering*, Vol. 2, p. 573, 2016.
52. A.J. Conway, P.J. Conway, and R.D. Fryar, Jr., Hand-actuated retention catheter, US Patent 5 360 402, assigned to Rochester Medical Corporation (Stewartville, MN), November 1, 1994.
53. F.E.B. Foley, Disposable cartridge for inflating bag catheters, US Patent 3 409 016, assigned to Selflate Corp., November 5, 1968.
54. A.J. Conway, P.J. Conway, and R.D. Fryar, Jr., Method of making catheter, US Patent 5 593 718, assigned to Rochester Medical Corporation (Stewartville, MN), August 2, 1994.
55. K. Sato and K. Senoo, Curable silicone rubber composition and measurement method for curable silicone rubber compositions, US Patent 9 115 268, assigned to Sumitomo Bakelite Co., Ltd. (Tokyo, JP), August 25, 2015.
56. J. Okada and K. Senoo, Silicone rubber-based hardening resin composition, molded article, and medical tube, US Patent 9 562 158, assigned to Sumitomo Bakelite Company Limited (Shinagawa-ku, JP), February 7, 2017.
57. T. Wada and K. Itoh, Heat-curable elastomeric silicone compositions, US Patent 3 671 480, assigned to Shin-Etsu Chemical Company (Tokyo, JP), June 20, 1972.
58. E.M. Jeram and R.A. Striker, High strength organopolysiloxane compositions, US Patent 3 884 866, assigned to General Electric Company (Waterford, NY), May 20, 1975.
59. W.J. Bobear, Inhibitor for platinum catalyzed silicone rubber compositions, US Patent 4 061 609, assigned to General Electric Company (Waterford, NY), December 6, 1977.
60. W.J. Bobear, Peroxide curing polysiloxane compositions having a high tear strength, US Patent 4 539 357, assigned to General Electric Company (Waterford, NY), September 3, 1985.

61. Japanese Standards Association, Rubber, vulcanized or thermoplastic – determination of tear strength – part 1: Trouser, angle and crescent test pieces, Online, electronic: https://infostore.saiglobal.com/en-gb/standards/jis-k-6252-1-2015-625073_SAIG_JSA_JSA_1435373/ JIS K 6252-1:2015, Japanese Standards Association, Tokyo, JP, 2015.
62. H.B. Kim, K.Y. Baik, M.H. Moon, and C.K. Sung, *Acta Physica Polonica A*, Vol. 129, p. 857, 2016.
63. J.L. Sommer and T.D. Brostrom, Suture sleeve, US Patent 7 082 337, assigned to Medtronic, Inc. (Minneapolis, MN), July 25, 2006.
64. J. Zhang, B.J. Ward, M.J. Tzivanis, X. Liu, A. Letailleur, and J.L. Ding, Silicone-based composition and article made therefrom, US Patent Application 20 180 171 143, assigned to Saint-Gobain Performance Plastics Corp., June 21, 2018.
65. Y. Yoon and R.h.S. Cho, Lens assembly including a silicone fresnel lens, US Patent Application 20 180 231 778, assigned to Facebook Technologies LLC, August 16, 2018.
66. C.-C. Kuo and J.-X. Lin, *The International Journal of Advanced Manufacturing Technology*, Vol. 99, November 2018.
67. A.V. Kaliyathan, A. Mathew, A.V. Rane, K. Kanny, and S. Thomas, Natural rubber and silicone rubber-based biomaterials in S. Thomas, P. Balakrishnan, and M.S. Sreekala, eds., *Fundamental Biomaterials: Polymers*, Woodhead Publishing Series in Biomaterials, chapter 4, pp. 71–84. Woodhead Publishing, 2018.
68. J.M. Courtney and T. Gilchrist, *Medical and Biological Engineering and Computing*, Vol. 18, p. 538, July 1980.
69. A. Rahimi and A. Mashak, *Plastics, Rubber and Composites*, Vol. 42, p. 223, 2013.
70. J. Curtis and A. Colas, Medical applications of silicones in B.D. Ratner, A.S. Hoffman, F.J. Schoen, and J.E. Lemons, eds., *Biomaterials Science*, chapter II.5.18, pp. 1106–1116. Academic Press, third edition, 2013.
71. C. Pinese, S. Jebors, P.E. Stoebner, V. Humblot, P. Verdié, L. Causse, X. Garric, H. Taillades, J. Martinez, A. Mehdi, and G. Subra, *Materials Today Chemistry*, Vol. 4, p. 73, 2017.
72. C.J. Boyer, J. Ambrose, S. Das, A. Humayun, D. Chappidi, R. Giorno, and D.K. Mills, *Medical Devices (Auckland, N.Z.)*, Vol. 11, p. 123, April 2018.
73. Z.-Y. Lei, T. Liu, W.-J. Li, X.-H. Shi, and D.-L. Fan, *International Journal of Nanomedicine*, Vol. 11, p. 5563, October 2016.
74. S.L. Evans and P.J. Gregson, *Materials Letters*, Vol. 16, p. 270 , 1993.
75. M. Matsuoka, T. Akasaka, T. Hashimoto, Y. Totsuka, and F. Watari, *Bio-medical Materials and Engineering*, Vol. 19, p. 155, 2009.
76. T. Ling, J. Lin, J. Tu, S. Liu, W. Weng, K. Cheng, H. Wang, P. Du, and G. Han, *Journal of Materials Science: Materials in Medicine*, Vol. 24, p. 2709, 2013.

77. C.V. Rahman, D. Ben-David, A. Dhillon, G. Kuhn, T.W.A. Gould, R. Müller, F.R.A.J. Rose, K.M. Shakesheff, and E. Livne, *Journal of Tissue Engineering and Regenerative Medicine*, Vol. 8, p. 59, 2014.
78. S.-L. Wang, X.-H. Shi, Z. Yang, Y.-M. Zhang, L.-R. Shen, Z.-Y. Lei, Z.-Q. Zhang, C. Cao, and D.-L. Fan, *PLOS One*, Vol. 9, p. e98320, 2014.
79. X. Zhou, X. Chen, T.-C. Mao, X. Li, X.-H. Shi, D.-L. Fan, and Y.-M. Zhang, *Plastic and Reconstructive Surgery*, Vol. 137, p. 690e, 2016.
80. L. Valentini, S. Bittolo Bon, and N.M. Pugno, *Advanced Functional Materials*, Vol. 27, p. 1606526, 2017.
81. D. Tamarkin, H. Shifrin, R. Keynan, E. Ziv, T. Berman, D. Schuz, and E. Gazal, Oil and liquid silicone foamable carriers and formulations, US Patent 9 549 898, assigned to Foamix Pharmaceuticals Ltd. (Rehovot, IL), January 24, 2017.

6

Other Uses

In this chapter, some special applications are collected that do not fit into the categories of automotive, electric, and medical.

6.1 Non-aqueous Organic Product Sensor

A detection system for the presence of a non-aqueous organic compound has been reported (1). The system contains a fiber-optic probe assembly with an optical fiber, which is susceptible to microbending anywhere along its length, and a body of material contained within a rigid containment structure. This body contains a material selected from the group consisting of a rubber, plastic and semicrystalline/rubbery polymeric material which is subject to a volumetric change capable of inducing a microbend in the optical fiber in the presence of a non-aqueous organic material.

Polymeric materials in contact with chemically and thermodynamically compatible organic liquids absorb and swell with the molecules of the organic liquid to form a swollen gel.

The swelling of a material in a liquid is commonly expressed in terms of parts per hundred (*pph*) relative to the initial dry weight of the material.

Since the polymer-polymer intermolecular forces are high because of, for example, crosslinking, crystallinity, complexation, strong hydrogen bonding or chain entanglement, the swollen polymer gel does not dissolve.

The swelling of the sensing polymer is largely dependent on the thermodynamic compatibility or match (lack of match) of their cohesive densities. The thermodynamic compatibility of a liquid and a polymer is often expressed empirically using solubility parameter theory where the solubility parameter is defined as the square root of the cohesive energy density. A match or near match of the solubility parameters of the target liquids and the sensor coating materials will, generally, indicate the potential sensing capability of the sensor cable.

The solubility parameter theory provides a useful guide for the selection of solvents for the application of polymeric coatings from organic solutions.

A suitable coating material is the rapid curing, solvent dispersible EP4412 Silastic Compound (Dow Corning).

The swelling behavior of a Silastic EP4412 silicone rubber film with a thickness of 200 μm in several hydrocarbon liquids is shown in Table 6.1 and the swelling behavior of a Silastic 9280/30E silicone rubber film is shown in Table 6.2.

Table 6.1 Swelling of a Silastic EP4412 silicone rubber film (1).

Time/	Uptake of Hydrocarbon Liquid /[*pph*]					Dodecyl
[*min*]	Kerosene	Petrol	Diesel	Gas Oil	Condensate	Benzene
5	154	191	74	60	182	52
10	160	192	81	68	190	61
20	159	196	81	68	182	65
30	162	191	83	69	190	65
60	165	199	82	69	194	66

Table 6.2 Swelling of a Silastic 9280/30E silicone rubber film (1).

Time/	Uptake of Hydrocarbon Liquid /[*pph*]					Dodecyl
[*min*]	Kerosene	Petrol	Diesel	Gas Oil	Condensate	Benzene
5	144	179	72	51	156	47
10	151	167	78	60	162	56
20	147	187	78	64	177	60
30	155	192	80	67	164	64
60	152	189	79	66	177	64
120	157	188	81	67	180	64

The data presented in Tables 6.1 and 6.2 demonstrate the potential for rapid response using materials in thin film form. The film samples were between 75# – 95# of their maximum swelling in the different investigated hydrocarbon liquids. The lower molecular weight, lower viscosity, more volatile liquids were observed to swell significantly in 10 s, such as kerosene, petrol, and condensate. The speed of response is proportional to the square of the thickness and so thicknesses of 0.50 μm and 0.100 μm will respond and swell more rapidly (1).

With respect to their mechanical strength in a swollen state, a qualitative assessment indicated that the EP4412 silicone rubber is significantly stronger than the 9280/30E silicone rubber. In comparison to other silicone rubbers, the combination of relative properties such as high hardness, high tensile strength, high elongation at break and high tear strength appear to indicate the significant retention of mechanical integrity in the swollen state.

The swelling of the polymeric materials in the presence of organic fluids produces a volumetric expansion that can be used to exert a mechanical force or pressure which in combination with an optical fiber system can be used for the detection of organic fluids. The extent of the mechanical response will be largely dependent on the degree to which the material absorbs and swells in the organic fluid and on the mechanical strength and/or modulus of the swollen gel. This response is reversible by evaporation of the organic liquid/vapor from the swollen material. According to the above-discussed properties, sensor cables were produced. The methods of fabrication has been detailed (1).

In summary, the principal advantage of the detailed sensor types is the ability to detect and locate a wide variety of organic fluids and vapors emanating from, for example, a fault in a pipeline or a storage tank (1).

6.2 Synthetic Leather

A method has been presented for manufacturing synthetic leather for various interior materials and various miscellaneous goods using liquid silicone rubber coating solutions (2). The physical properties ware optimized so as to be suitable for the use of different textiles.

The liquid silicone rubber used may be composed of an A part and a B part, as shown in Table 6.3.

Table 6.3 Liquid silicone rubber composition (2).

Compound	Amount/[%]
Part A	
Dimethylvinyl-terminated dimethyl siloxane	30–60
Trimethylated silica	<10
Dimethylvinylated and trimethylated silica	<10
Alumina hydrate	30–60
Platinum Catalyst	<1
Part B	
Dimethylvinyl-terminated dimethyl siloxane	30–60
Dimethyl, methylhydrogen siloxane	30–60
Trimethylated silica	<10
Dimethylvinylated and trimethylated silica	<1
Alumina hydrate	<10

Two parts may be mixed with each other with similar contents to each other. Furthermore, the A and B parts may be mixed with different contents and at the same time a content of a curing agent may be adjusted. Even in the case of adjusting a mixing ratio, basic addition reaction patterns are equal, and in order to improve processability, 10 to 300 parts by weight of a diluent according to the properties of a final product may be added based on 100 parts by weight of the liquid silicone rubber.

Since it is intended to provide a synthetic leather product used as an interior material of an aircraft, an interior material of a furniture, and an interior material of a vehicle and a synthetic leather product for a bag, miscellaneous goods, or the like, it is important to select a type of silicone rubber with suitable properties (2).

6.3 Two-Part Curable Composition

In silicone rubber compositions, which contain a calcium carbonate powder and which are cured by a hydrosilylation reaction, the calcium carbonate powder contains alkaline components as impurities, and therefore involve problems such as hydrogen gas being generated during storage due to reactions with organopolysiloxanes having silicon-bonded hydrogen atoms (3). So, storing an addition reaction curable silicone rubber composition as two components is commonly used.

A two-part curable liquid silicone rubber composition has been described that is composed of two compositions that are stored separately, in which the physical properties and the adhesion to silicone rubber of a cured product thereof do not decline when mixed, even in cases where a mixture ratio of the first composition to the second composition varies (3).

The methods of preparation of the two compositions have been described as follows: First composition (3):

Preparation 6–1: The first liquid composition was prepared by mixing 98.0 parts by mass of a dimethylpolysiloxane capped at both molecular terminals with dimethylvinylsiloxy groups and having a viscosity of 40,000 *mPa s*, 2.0 parts by mass of a dimethylpolysiloxane having vinyl groups at both molecular terminals and on the side molecular chains and having a viscosity of 350 *mPa s* (vinyl group content: 0.93% by mass, average number of vinyl groups on the molecular side chains: 2.5), 3.0 parts by mass of fumed silica surface-treated with dimethyldichlorosilane and having a BET specific surface area of about 110 m^2g^{-1}, 25 parts by mass of precipitated calcium carbonate powder surface-treated with fatty acids and having a BET specific surface area of 18 m^2g^{-1} (Hakuenka CCR, manufactured by Shiraishi Kogyo Kaisha, Ltd.), 3.0 parts by mass of dimethylpolysiloxane capped at both molecular terminals with silanol groups and having a viscosity of 40 *mPa s*, a 1,3-divinyltetramethyl disiloxane solution of a 1,3-divinyltetramethyl disiloxane platinum complex (included in an amount such that the amount of platinum metal in the catalyst is 50 parts by mass per 1,000,000 parts by mass of the present composition), and 2.0 parts by mass of a pigment paste in which 40 parts by mass of red iron oxide are compounded with 60 parts by mass of a dimethylpolysiloxane capped at both molecular terminals with dimethylvinylsiloxy groups and having a viscosity of 10,000 *mPa s*.

Second composition (3):

Preparation 6–2: The second liquid composition was prepared by mixing 98.9 parts by mass of a dimethylpolysiloxane capped at both molecular terminals with dimethylvinylsiloxy groups and having a viscosity of 40,000 *mPa s*, 3.0 parts by mass of fumed silica surface-treated with dimethyldichlorosilane and having a BET specific surface area of approximately 110 m^2g^{-1}, 30 parts by mass of a precipitated calcium carbonate powder surface-treated with a fatty acid and having a BET specific surface area of 18 m^2g^{-1} (Hakuenka CCR manufactured by Shiraishi Kogyo

Kaisha, Ltd.), 6.0 parts by mass of a dimethylpolysiloxane capped at both molecular terminals with silanol groups and having a viscosity of 40 *mPa s*, 1.05 parts by mass of a copolymer of dimethyl siloxane and methylhydrogensiloxane capped at both molecular terminals with trimethylsiloxy groups, which had a kinetic viscosity of 21 mm^2s^{-1}, had an average of 1.9 silicon-bonded hydrogen atoms in a molecule in side chains and had a silicon-bonded hydrogen atom content of approximately 0.098% by mass (at an amount where the molar ratio of silicon-bonded hydrogen atoms in the mixture relative to silicon-bonded vinyl groups contained in the total composition was 0.10), and 5.10 parts by mass of a dimethylpolysiloxane capped at both molecular terminals with dimethylhydrogensiloxy groups, which had a kinetic viscosity of 10 mm^2s^{-1} and a silicon-bonded hydrogen atom content of approximately 0.016% by mass (at an amount where the molar ratio of silicon-bonded hydrogen atoms in the mixture relative to silicon-bonded vinyl groups contained in the total composition was 0.90).

The silicone rubber compositions were mixed and prepared such that mass ratios of the first liquid composition to the second liquid composition were 110:100, 100:100, and 100:110. The physical properties and adhesive strength and adhesion ratio of the silicone rubber obtained by curing this silicone rubber composition were measured and the results thereof are recorded in Table 6.4.

Table 6.4 Properties dependent on mixing ratio (3).

Mixing Ratio	Hardness Asker C strength	Tensile strength/[*MPa*]	Elongation /[&]
110/100	24	0.14	4600
100/100	27	0.16	4300
100/110	24	0.22	3600

As shown in Table 6.4, no significant changes in the physical properties were seen at mixing ratios of 110:100, 100:100 and 100:110 (3).

6.4 Microchannel Thermocured Silicone Rubber

Porous polymers (4) can be used in the field of membranes and implants in medicine (5) and fuel cell technology (6).

Here, silicones are of special interest, due to their unique properties such as transparency, biocompatibility, high elasticity and thermal stability, low glass transition temperature and high gas permeability.

There are several methods of forming porous structures. One of them consists of obtaining a polymer filled with microparticles of water-soluble inorganic salts. Washing with water leads to the formation of pores. The creation of microchannels can be done by photolithography (7), capillary force lithography (8), mold and hot embossing (9, 10), sputtering on a sacrificial polymer fiber (11), and hot roller embossing (12). These procedures provide products with a strictly set network of microchannels.

Whiskers are a special form of crystal compounds (13–15). Natural and artificially obtained whiskers have attracted attention due to their high hardness.

Silicone rubber penetrated by a network of microchannels with diameters 10–600 μm and lengths of 2–30 mm could be formed by washing away of whiskers prepared from *p*-aminobenzoic acid, c.f. Figure 6.1, from a thermocured poly(methylphenyl siloxane) composition (16).

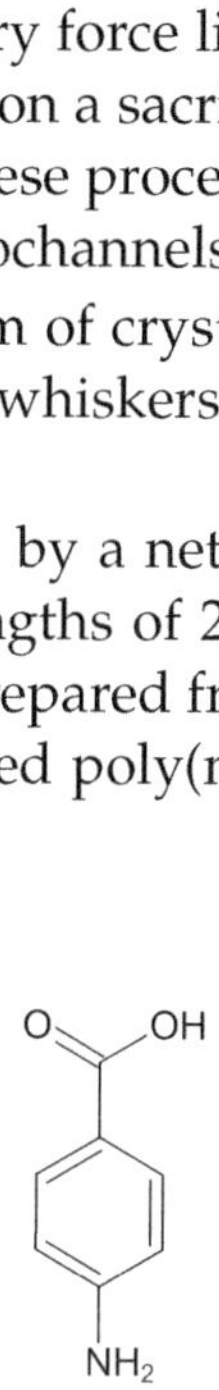

Figure 6.1 *p*-Aminobenzoic acid.

6.5 Dry Cleaning of Surfaces

From esthetic, hygiene, and health viewpoints, thorough house cleaning is an important and demanding task. The removal of household dust is particularly challenging because of its constant formation and extensive distribution. Household dust consists of a multitude of organic and inorganic particles, fibers and substances which have their origin in animals (domestic animals, mites), plants (pollen, seeds), people (flakes of shedded skin, hair), sand, molds, bacteria, viruses, household chemicals (softeners), and others, and thus can have harmful health effects (e.g., allergies).

A multitude of cleaning agents, devices and utensils are available for the collection and removal of loose dirt particles in the domestic sector. Basically, a distinction can be made between dry and wet-dry cleaning methods. For example, vacuum cleaners can be used for the dry cleaning of large, freely accessible surfaces. For the dry cleaning of smaller, in particular curvilinear, poorly accessible hard surfaces (furniture, standard lamps, pictures, televisions, radiators, ornamental objects, etc.) the use of various dusters, feather dusters, brooms, hand brushes, wipers, brushes, and mops, whose mode of operation essentially is based on loose adsorption of the dirt particles on the fabric or the fibers or bristles of the cleaning device, have proven to be effective.

For the dry cleaning of soft surfaces, in particular textiles, brushes and lint rollers can, for example, be used. The effectiveness of cleaning devices based on dry physisorption of the dirt particles is predominantly determined by two factors.

First, by the size of the surface of the cleaning device collecting the dirt particles, and second, by the strength of the adsorption forces acting between dirt particles and the surface itself. The effective surface of the cleaning device can, for example, be considerably increased by the use of special fabric structures, very thin fibers, and bristles (e.g., with slit ends or split), which are, for example, exploited in microfiber cloths, feather dusters, dust wipers, etc. Increasing the adsorption forces of the cleaning device is of particular importance. If the adsorption forces are too small, the dirt particles can only inadequately be bound by the cleaning device, which has the result that they are only distributed from one place to the other, but not removed.

An increase in the adsorption forces can be achieved by stronger adhesiveness of the surface as described in the literature (17).

In order to avoid direct contact of the adhesive surface, e.g., of a lint roller with the substrate to be cleaned, the dirt particles taken up and adsorbed from a first non-adhesive roller can be transferred to a second adhesive roller, described elsewhere (18)

However, this method has the disadvantage that simple cleaning of the cleaning device by simply shaking out, knocking out or wiping is now no longer possible and the surface (e.g., of a roller) saturated with dirt particles must be disposed of (19). Furthermore, a sticky surface can also contaminate or damage the object to be cleaned. Finally, this method fails with fabric-type, fiber- or bristle-based cleaning devices since the fibers adhere together and the high coefficient of sliding friction of the cleaning device makes sliding on the object to be cleaned impossible.

A further possibility for increasing the adsorptivity consists of impregnation or coating of the fibers of the cleaning device, for which oils, plant waxes, olive leaf extracts, etc., are used. These parts must, as a rule, be thrown away after use. Another important possibility for increasing the adsorption forces consists in binding the dirt particles electrostatically to a surface of the cleaning device. The electrostatic interaction between dirt particles and cleaning device is a universal process, which is always induced by friction or contact with different materials. This effect, also described as triboelectric charging, occurs to a different degree of intensity depending on material combination, intensity of friction or contact, atmospheric humidity, temperature, etc.

While, for example, lamb's wool or ostrich feathers scarcely become electrostatically charged through friction, a poly(urethane) foam achieves strong positive charging, and a Teflon™ surface achieves strong negative electrostatic charging. More precise consideration shows that different materials in general also possess different electrochemical potentials, which has the result that during the friction process or even simple contact of two materials, electrons are transferred from the material with the higher electrochemical potential to the material with the lower electrochemical potential, as a result of which the former undergoes a positive electrostatic charging, and the latter a negative electrostatic charging.

Since the cleaning devices based on dry cleaning considered here come into intensive (frictional) contact with the object to be cleaned, it can be assumed that triboelectric effects contribute not inconsiderably to the adsorption strength of the cleaning device, particularly since the cleaning devices are mostly of plastic materials, i.e., electrical insulators. The surface of the cleaning device electrostatically charged as a result of friction or contact is able not only to attract oppositely charged dirt particles, but also dirt particles which are polar or polarized by induction, which themselves have no net charge (19).

If two different materials are rubbed together or separated after pressing together, an electrostatic charging takes place, whereby one of the materials becomes negatively and the other positively charged. Different materials have a different electron affinity, because of which the material with the higher electron affinity undergoes a negative charging, and that with lower electron affinity a positive charging. Furthermore, upon rubbing two materials together, it is found that the strength of the charging, i.e., the quantity of charge transferred, is different depending on the material combination. In accordance with the strength and the sign (positive or negative) of the charging in a frictional process, the materials can be arranged in the so-called triboelectric series. The materials with the strongest tendency to positive charging form the positive end of the series, and those with the strongest tendency to negative charging form the negative end. Consequently, materials which show a very low to absolutely no tendency to triboelectric charging are situated in the middle part of this series. The strength of the tendency for positive or negative charging of a material can be measured by the quantity of positive or negative charge (in coulombs) transferred onto the material per Joule frictional energy under defined conditions and thus be quantified (in coulombs/joule). The greater the difference in the tendencies of two materials to charging (i.e., the further apart from one another these are in the triboelectric series), the more strongly the triboelectric charging will take place during the rubbing process.

Silicone polymers are among the material with the highest tendency for negative charging. They are thus located at the negative end of the triboelectric series. If, therefore, a silicone polymer is rubbed with another material, the silicone polymer always becomes negatively charged and the other material positively, and this becomes more pronounced the further away the other material is located from the silicone polymer in the triboelectric series. However, in the use of silicone polymers for the adsorption of particles in cleaning devices, the comparatively high price of the silicone elastomer compositions is an obstacle. Secondly, the production of textile fabrics and fibers from silicone polymer would be extremely laborious and thus expensive. Finally, the mechanical properties and frictional properties of silicone polymers are unsuitable for the functionality of cleaning devices.

Here, a simple method was developed, which makes it possible to increase the effectiveness of cleaning devices based on dry physisorption, without adversely altering their specific functionality (19).

A 2 *mm* thick sheet of liquid silicone rubber was produced. For this, two components of the liquid silicone rubber ELASTOSIL™ LR 3003/50 (Wacker Chemie AG) in the mass ratio 1:1 were intimately mixed on a roller mill for 15 *min* at a roller temperature of 25°C. This mixture was then crosslinked in a hydraulic press for 15 *min* at a temperature of 170°C. The demolded 2 *mm* thick silicone elastomer sheet was subjected to a tempering lasting 4 *h* at 200°C in an air circulation drying cabinet.

As materials used for rubbing the plastic plates, a Microfasertuch Premium Professional microfiber cloth from MOPPTEX GmbH (Austria), a silicone-free cotton cloth: standard cotton fabric, and a filter paper from Schleicher & Schuell, Dassel (Germany), were tested.

A silicone content of less than 1% already causes a drastic increase in the field strength, which corresponds to a stronger electrostatic charging of the plastic as a result of friction (19).

6.6 Adhesive Tapes

Known pressure-sensitive adhesives include adhesives based on silicones (20). Silicone adhesives are superior to acrylic adhesives and rubber adhesives in properties such as heat resistance, cold resistance, weather resistance, chemical resistance, and electrical insulation, and adhere well to various adherends. Silicone adhesives are thus used in various adhesive tapes such as heat-resistant tapes, electrically insulating tapes, heat sealing tapes, and plating masking tapes. When an adhesive tape is used in an application that requires heat resistance, a base with excellent heat resistance, such as a fluororesin base or a glass cloth base impregnated with a fluororesin, is often used with the silicone adhesives. Silicone adhesives include peroxide-curable adhesives and addition reaction-curable adhesives.

To increase the heat resistance of an adhesive tape, not only the properties of a base but also those of a silicone adhesive need to be improved. In view of applications of the adhesive tape, it is desired to improve, in particular, the ability to maintain the adhesive strength in high-temperature environments among the properties of the silicone adhesive.

A silicone adhesive composition has been described that exhibits a high load-bearing capacity even in higher-temperature environments and an adhesive tape that exhibits high load-bearing capacity and good shelf life even in such high-temperature environments (20).

A schematic of an adhesive tape is shown in Figure 6.2. The tape includes a base layer and an adhesive layer placed on the base layer; the adhesive layer is formed from a silicone adhesive composition. The adhesive layer is a cured layer of the silicone adhesive composition.

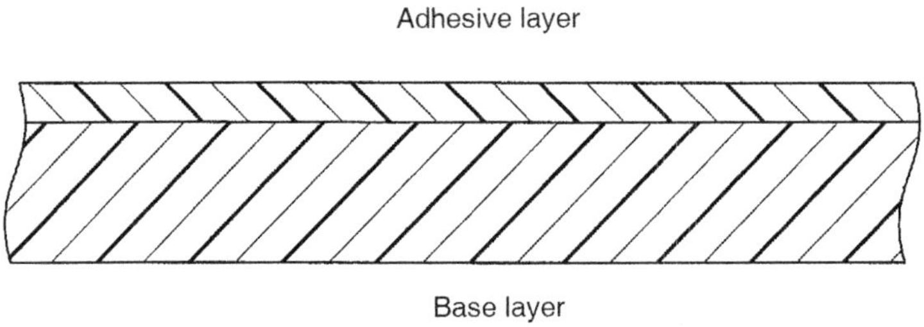

Figure 6.2 Adhesive tape (20).

The method for producing such an adhesive tape is done by heating an original tape, including a base and the above silicone adhesive composition placed on the base, and thereby forming the composition into an adhesive layer so as to obtain an adhesive tape including the base and the adhesive layer placed on the base.

So it is possible to obtain a silicone adhesive tape that exhibits a high load-bearing capacity and a good shelf life even in such high-temperature environments. The preparation of an adhesive tape runs as follows (20):

Preparation 6–3: First, 100 parts by weight of dimethylpolysiloxane (M_w 700 *kD*, raw rubber-like form) were mixed with 170 parts by weight of an MQ resin as a silicone resin (M_w 5.5 *kD*), the MQ resin consisting of 44 *mol*% of $(CH_3)_3SiO_{1/2}$ units corresponding to the M units and 56 *mol*% of SiO_2 units corresponding to the Q units, and toluene.

Then, the resulting mixture was held at 100°C to 120°C for 4 *h* to allow the dimethylpolysiloxane and the MQ resin to undergo partial condensation. After that, toluene was further added to adjust the content of solids (non-volatile components). This yielded a toluene solution A (concentration of 60%) of the dimethylpolysiloxane, the MQ resin, and partial condensates of the dimethylpolysiloxane and the MQ resin. The solution A obtained was viscous, colorless, and transparent.

Next, 100 parts by weight of the obtained toluene solution A, 0.96 parts by weight (in terms of solid content) of a xylene solution of dibenzoyl peroxide as an organic peroxide curing agent (NYPER BMT-K40, manufactured by NOF Corporation and having a solid content of 40%), and 5 parts by weight of RTV 4086 (manufactured by Dow Corning Toray Co., Ltd.) as an addition reaction-curable silicone rubber, were mixed. Further, the concentration was adjusted with toluene, and thus a silicone adhesive composition having a concentration of 40% was obtained.

Then, the silicone adhesive composition obtained was applied to a polyimide film (Kapton 10011, manufactured by Du Pont-Toray Co., Ltd.) serving as a base, and a coating layer with a thickness of 40 μm was thus formed. Subsequently, the base and the coating layer were heated together using a drying machine held at 200°C for 3 *min* to cure the coating layer and thus obtain an adhesive tape including the base and an adhesive layer placed on a surface of the base, the adhesive layer being formed from the applied adhesive composition.

6.7 Capsules for Beverages

Devices are used for preparing a beverage by injecting a pressurized fluid into a capsule, especially for producing coffee or coffee type beverages. In addition, other comestible ingredients, such as chocolate or milk products, can be contained within the capsule (21).

By means of an interaction of these ingredients with a liquid, a beverage or other comestibles, such as soup, can be produced. The interaction can be, for example, an extraction, brewing, dissolution, etc., process. Such a capsule is particularly adapted to contain ground coffee in order to produce a coffee beverage by having hot water under pressure enter the capsule and draining a coffee beverage from the capsule.

The advantages of such a system are, in particular, the conservation and freshness of the ingredients, as well as the possibility of facilitating the operations of preparing the beverage.

Systems and methods for obtaining fluid comestibles from substance-containing capsules are known in the art (22). The capsule 101, as shown in Figure 6.3, has a frustoconically shaped cup 102 which may be filled, e.g., with roasted and ground coffee 103, and which is closed by a foil-like tear face cover 104 welded and/or crimped to a flange-like rim which extends laterally from the side-wall of the cup 102. A capsule holder 111 is composed of a flow grill 112 with relief surface element members 113.

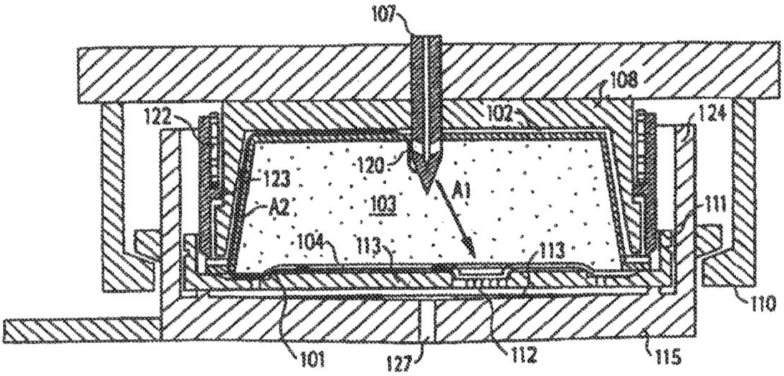

Figure 6.3 Capsule schematic (21).

The capsule holder 111 is accommodated in its support 115, which has a lateral wall 124 and a bore 127 for the passage of extracted coffee beverage. The extraction system further contains a water injector 107 having a water inlet channel 120 and an annular element 108 with an internal recess of which the shape substantially corresponds to the outer shape of the capsule. On its outer part, the annular member 108 comprises a spring 122 holding a ring 123 for releasing the capsule on completion of extraction.

In operation, a capsule 101 is placed in the capsule holder 111. The water injector 107 perforates the upper face of the cup 102. The lower tear face 104 of the capsule rests on the radially arranged members 113 of the capsule holder 111.

The water is injected through the channel 120 of the water injector 107 and impinges on the bed 103 of coffee. The pressure in the capsule 101 increases and the tear face 104 increasingly follows the shape of the radial opening relief members 113. Such radial opening relief members could be replaced by pyramid-shaped relief members or other shapes of relief. When the constituent material of the tear face reaches its breaking stress, the tear face tears along the relief members. The extracted coffee flows through the orifices of the flow grill 112 and is recovered in a container (not shown) beneath the bore 127.

A method has been proposed for providing a rubber-elastic sealing member on a capsule, wherein the capsule has a cylindrical or frustoconical body portion having a side wall and a flange-like rim portion, wherein the rim portion extends from the side wall of the body portion. The method consists of the steps of (21):

1. Applying an uncured sealing composition in a liquid or viscous state onto the rim portion of the capsule,
2. Heating the applied sealing composition such that the sealing composition at least partially migrates towards the side wall until it reaches the side wall, and
3. Curing the sealing composition after the migration towards the side wall.

According to the method, the sealing composition can be accurately and conveniently placed at the intersection of the side wall and the flange-like rim portion of the body of the capsule such that a water-tight sealing is obtained at said intersection.

Preferably, the capsule is made of metal such as aluminum and formed by a deep drawing operation. Hence, the body of the capsule may be covered with a lacquer and a residue of lubricant necessary during deep drawing of the body, which may promote the attraction of the sealing composition to the body of the capsule. Accordingly, the heating of the sealing composition before the curing process enables a desired equal distribution of the uncured sealing composition at the intersection between the rim portion and the body portion of the capsule.

The sealing composition is preferably a resilient, rubber-elastic material which is selected from elastomers, silicones, plastics, latex, balata, or others. The sealing composition is preferably transparent. Preferably, a silicone rubber marketed as Addisil™ 1540D commercialized by General Electric Company can be used as liquid silicone rubber for the sealing composition (21).

The migration of the sealing composition towards the side wall of the capsule is preferably obtained by heating of the sealing composition at an oven temperature of 60°C to 80°C for 1 to 4 *min*. After the curing process, a stable and resilient sealing means is formed at the capsule (21).

6.8 Usage for Toner

Copiers and electrophotographic printers use colored particles referred to as toners. Single-component toners predominantly in use today are made from polyester resins and styrene-acrylic resin. These toners are required to be quick melting because of the higher printing speeds. In addition, from the standpoint of reducing energy consumption by the machine itself, the trend in toner design melting points is toward lower temperatures (23).

At the same time, to enable the development roll to handle higher printing speeds, lower hardness, and increased surface smoothness have been required of the rubber. However, with the lower toner melting points, the influence on the toner of frictional heat generated on the development roll has become larger, making low-temperature control of the development roll surface important.

So, there is a need to provide a silicone rubber composition for a silicone rubber development member, which is characterized by excellent image properties and high thermal conductivity, and to provide a thermally conductive silicone rubber development member, such as a roll or belt, having a silicone rubber layer obtained by curing such a composition.

An addition-curable or organic peroxide-curable silicone rubber composition has been developed for a thermally conductive silicone rubber development member such as a silicone development roll or a silicone development belt (23). The composition has a silicone rubber layer obtained by curing of such a composition. The silicone rubber layer is obtained by curing this silicone rubber composition to which has been added a silicon metal powder and carbon black. This efficiently lowers the surface temperature of the development roll or development belt and is thereby able to reduce damage to the toner.

The silicone rubber composition for a thermally conductive silicone rubber development member contains (23):

1. An organopolysiloxane containing in the molecule at least two alkenyl groups which bond with silicon atoms,
2. A thermally conductive powder having an average primary particle size of not more than 30 μm and a thermal conductivity of at least 10 $W m^{-1} K^{-1}$,

3. Carbon black, and
4. A curing agent capable of curing component 1.

The structure of organopolysiloxane is basically a linear structure in which the backbone is made up of recurring diorgano siloxane units, such as dimethyl siloxane units, diphenyl siloxane units, methylphenyl siloxane units, methyltrifluoropropyl siloxane units or vinylmethyl siloxane units, and both ends of the molecular chain are capped with triorganosiloxy groups, such as trimethylsiloxy, vinyldimethylsiloxy, divinylmethylsiloxy, trivinylsiloxy, vinyldiphenylsiloxy, vinylmethylphenylsiloxy, phenyldimethylsiloxy or diphenylmethylsiloxy groups, although the structure may be partially composed of branched structures, cyclic structures and the like.

The thermally conductive powder is used for imparting thermal conductivity to the silicone rubber composition. Illustrative examples of the thermally conductive powder include thermally conductive inorganic powders such as silicon metal powder, alumina, aluminum, silicon carbide, silicon nitride, magnesium oxide, magnesium carbonate, zinc oxide, aluminum nitride, graphite, and fibrous graphite. Most preferably silicon metal powder can be used. Silicon metal has a good thermal conductivity, in addition to which it has a low Mohs hardness. Moreover, because silicon metal readily shatters when struck and has a low ductility, one property of the metal powder itself is that, even when subjected to high shear, it does not easily agglomerate. The thermally conductive powder has an average primary particle size of not more than 30 μm. The specific thermally conductive powder is compounded with the organopolysiloxane.

The organic peroxide curing agent may be one that is used as a catalyst to promote crosslinking reactions on the organopolysiloxane. Illustrative examples include, dibenzoyl peroxide, 2,4-dichlorobenzoyl peroxide, *p*-methylbenzoyl peroxide, *o*-methylbenzoyl peroxide, dicumyl peroxide, 2,5-dimethylbis(2,5-*tert*-butylperoxy)hexane, di-*tert*-butyl peroxide, *tert*-butyl perbenzoate and 1,1-bis(*tert*-butylperoxycarboxy)hexane. Some peroxides are shown in Figure 6.4.

The resulting silicone rubber composition for the thermally conductive silicone rubber development member can be molded for

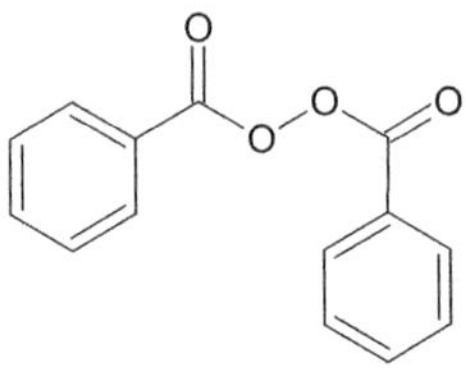

Dibenzoyl peroxide

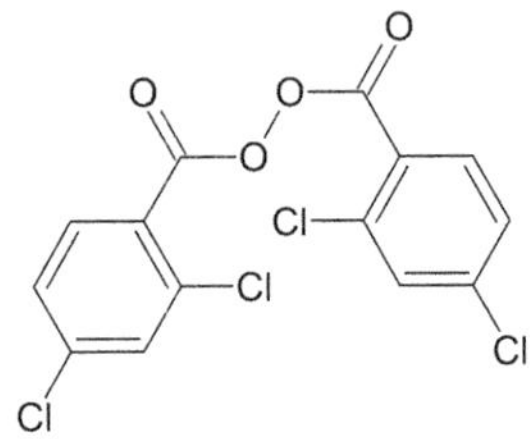

2,4-Dichlorobenzoyl peroxide

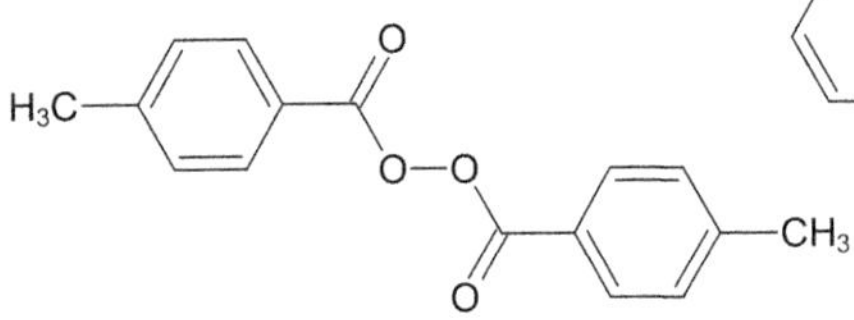

p-Methylbenzoyl peroxide

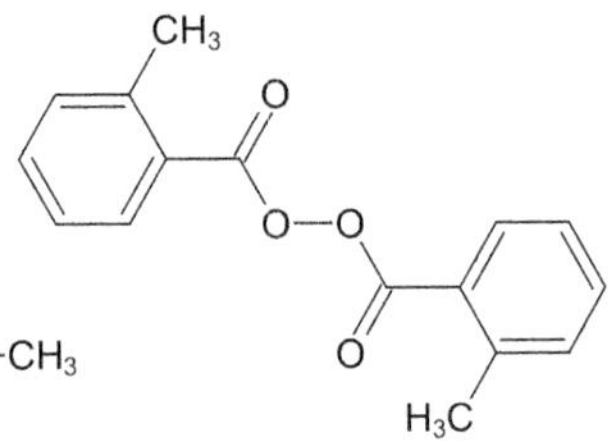

o-Methylbenzoyl peroxide

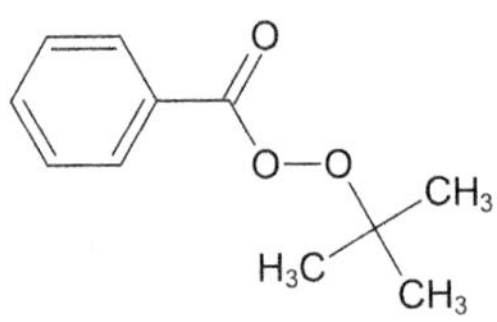

tert-Butyl perbenzoate

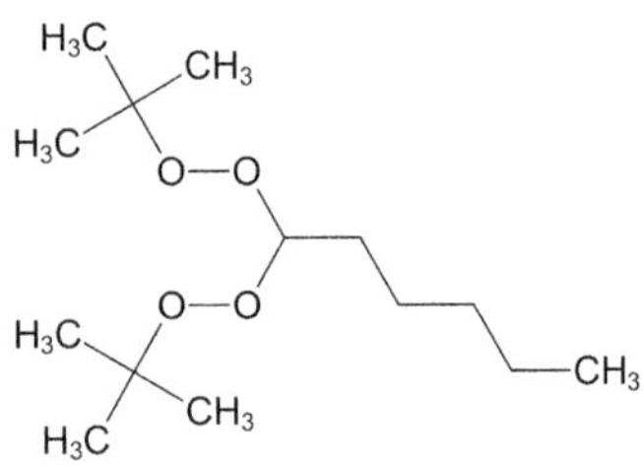

1,1-Bis(*tert*-butylperoxycarboxy) hexane

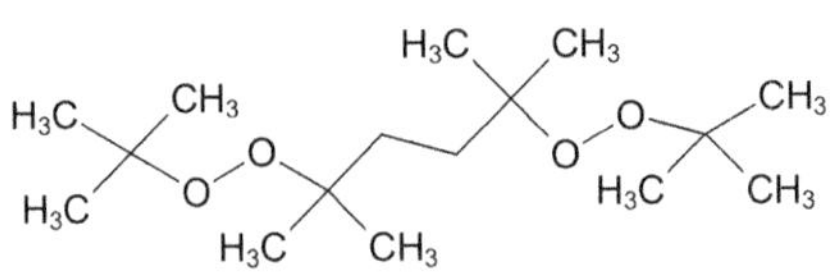

2,5-Dimethylbis(2,5-*tert*-butylperoxy)hexane

Figure 6.4 Peroxides.

the required application by any of the various molding methods commonly used for molding silicone such as casting, liquid injection molding, and pressure molding. The molding conditions are preferably in the range of 70°C to 400°C for a period of from several seconds to one hour. In cases where secondary vulcanization is carried out after molding, such secondary vulcanization is preferably carried out in the range of 150°C to 250°C for a period of from 1 *h* to 30 *h*.

In the development roll, a thermally conductive cured layer of the silicone rubber composition (silicone rubber layer) is formed on an outer peripheral surface of a core bar. In this case, the material, dimensions and the like of the core bar may be suitably selected according to the type of roll, although the core bar is typically made of, for example, aluminum, iron or stainless steel. It is preferable for the surfaces of these core bars to be treated with a primer such as a silane coupling agent or a silicone adhesive so as to further strengthen adhesion with the silicone rubber layer.

A urethane resin layer, silicone-modified urethane resin layer, or silane coupling coat may additionally be formed on the outer periphery of the silicone rubber layer. Here, the urethane resin is exemplified by resins obtained by reacting a polyether polyol or a polyester polyol with an aromatic polyisocyanate or an aliphatic polyisocyanate. The silicone-modified urethane resins can be obtained by curing a polyol or polyisocyanate in which a portion of the main chain or side chains has been modified with silicone units. An example preparation is as follows (23):

Preparation 6–4: A planetary mixer was charged with 60 parts by weight of a linear dimethylpolysiloxane capped at both ends of the molecular chain with dimethylvinylsiloxy groups (degree of polymerization, 500), 1.0 part by weight of hydrophobized fumed silica having a BET specific surface area of 110 m^2g^{-1} (R-972, from Nippon Aerosil Co., Ltd.), 4.0 parts by weight of Denka Black powder (Denki Kagaku Kogyo KK; average primary particle size, 40 *nm*), which is an acetylene black-type of carbon black, and 70 parts by weight of ground silicon metal powder A (average primary particle size, 5 μm), and stirring was carried out a room temperature of 23°C for 2 *h*.

The mixture was applied to a three-roll mill and dispersion was carried out. The mixture was then returned to the planetary mixer, where 40 parts by weight of a linear dimethylpolysiloxane capped at both ends of the molecular chain with trimethylsiloxy groups and having methylvinyl siloxane units on the main chain and pendant vinyl groups (degree of polymerization, 300; vinyl group content, 0.000075 $mol\,g^{-1}$), 1.0 part by weight of a methylhydrogenpolysiloxane having SiH groups at both ends and on side chains (degree of polymerization, 17; SiH group content, 0.0038 $mol\,g^{-1}$); a dimethyl siloxane-methylhydrogensiloxane copolymer capped at both ends of the molecular chain with dimethylhydrogensiloxy groups), 0.05 part by weight of ethynylcyclohexanol and 0.05 part by weight of tetramethyltetravinylcyclotetrasiloxane as reaction regulators, and 0.1 part by weight of platinum catalyst (Pt concentration, 1%) were added and stirring was continued for 15 minutes, thereby preparing an addition-curable, electrically conductive, liquid silicone rubber composition. The compounds used here are shown in Figure 6.5.

Ethynylcyclohexanol Tetramethyltetravinylcyclotetrasiloxane

Figure 6.5 Reaction regulators (23).

The resulting addition-curable, electrically conductive, liquid silicone rubber composition was liquid injection-molded over a 10 *mm* diameter core bar in a casting mold having a mold inside diameter of 16 *mm*, and cured by 20 *min* of heating at 120°C. This molding was polished, thereby forming a development roll 1 having an outside diameter of 14 *mm*, a rubber layer thickness of 2 *mm* and a rubber length of 220 *mm*.

The addition-curable, electrically conductive, liquid silicone rubber composition and development roll 1 thus obtained were subjected to various evaluations by the measurement methods. It was found that the silicone rubber composition for a thermally conductive silicone rubber development member was characterized by having an excellent heat dissipating ability, high elasticity, and low hardness, in addition to which the roll appearance was good.

6.9 Acoustic Applications

Exposure to excessive or repetitive noise is a potential risk especially for workers employed in the heavy engineering industry, including thermal and hydro power plants. Working in such conditions over a long period of time can result in hearing loss and other problems (24).

Noise reduction by porous materials prepared from metals, polymers, and ceramics has been a subject of research (25–28).

In comparison to porous materials prepared from metals and ceramics, polymers are easy to use as claddings because they are flexible, easy to manufacture, and less expensive. Also, empirical models have been developed that demonstrate the noise-attenuating capability of porous materials, especially at medium and high frequencies (29–31).

The acoustic properties of porous materials can be tailored by balancing the porosity, density, and thickness.

A comprehensive and scalable acoustic material preparation has been reported using a medium internal phase emulsion-templating technique from a single-part room temperature vulcanizable silicone rubber and pore generator, i.e., a water and mineral oil mixture (24).

A poly(dimethyl siloxane) variant, i.e., Powersil 567 from Wacker Silicones, which is one-component room temperature vulcanizable (RTV-1) liquid silicone rubber, was used as the matrix for preparing the sound-absorbing material. The reason for choosing LSR, a dielectric material, as the polymer matrix is its compatibility and applicability in the electrical industry.

By optimizing the mixing conditions, a pore-generator-volume and cure-schedule macroporous sound-absorbing silicone rubber was developed.

By optical and scanning electron microscopy, the pore size, their interconnectivity, and heterogeneity were analyzed. For an optimized sample (0.5 volume ratio water and 3.67 volume % mineral oil, porosity 83%, thickness ~ 12.5 ± 1 *mm*) the average sound absorption coefficient and transmission loss measured by an impedance tube technique in the 50–6400 *Hz* range were 0.45 and 6.7–42 *dB*, respectively. Insertion loss measurements in the 125–6300 *Hz* range with the material as cladding inside (100%) and outside (85%) of an acoustical enclosure demonstrated minimum/maximum noise reduction of 2.31 *dB* at 630 *Hz*/20.9 *dB* at 4000 *Hz* and 3.55 *dB* at 400 *Hz*/13.71 *dB* at 2000 *Hz*, respectively (24).

6.10 High Temperature Gas Line Heater System

Many industries, such as the telecommunication, chemical, and pharmaceutical industries, require heated conduit lines for fluids used in manufacturing (32). The microelectronic semiconductor fabrication industry in particular has the need for highly regulated gas transmission lines leading to and from the fabrication vessels used to process semiconductor wafers. These specialized gases are used for processing applications such as low pressure chemical vapor deposition (LPCVD), metal organic chemical vapor deposition (MOCVD), high density chemical vapor deposition (HDCVD), atomic layer deposition (ALD), and plasma etch.

A semi-flexible laminar silicone foam rubber-base heater for high temperature gas supply and exhaust lines used in microelectronic semiconductor fabrication uses superimposed adhesively bonded layers of an etched foil heater element containing layer, a high--to-low heat thermal transition layer, a thermal fuse laden silicone foam rubber insulation layer and a durable wrapping layer (32).

The system can provide for the combined use of highly controllable etched foil heater elements, clean silicone foam rubber insulation, and inexpensive mechanical thermal fuses safely in a high heat gas supply and exhaust line environment.

Referring to Figure 6.6, each cover piece 3 can include a number of stacked, lamellarly successive layers including a heater element containing layer 55 located proximate to the gas line, a thermal transition layer 70, a silicone foam rubber insulation layer 90, and an outer coating or wrap layer 110. Each layer can be located radially adjacent or radially spaced apart from other layers. Some layers, such as the heater element containing layer, can be made up of stacked, lamellarly successive sublayers, including, for example, the element itself and insulating and bonding layers. The layers and sublayers can be bonded to adjacent layers or sublayers by a bonding material layer. For clarity, there are no bonding layers shown in Figure 6.6.

In the case of cylindrical transmission line the layers can be coaxially and substantially cylindrically spaced apart from an inner location adjacent to the outer side wall of the gas line to an outer location on the outer surface of the heater piece. The term inner is used to refer to a location relatively closer to the gas line while the

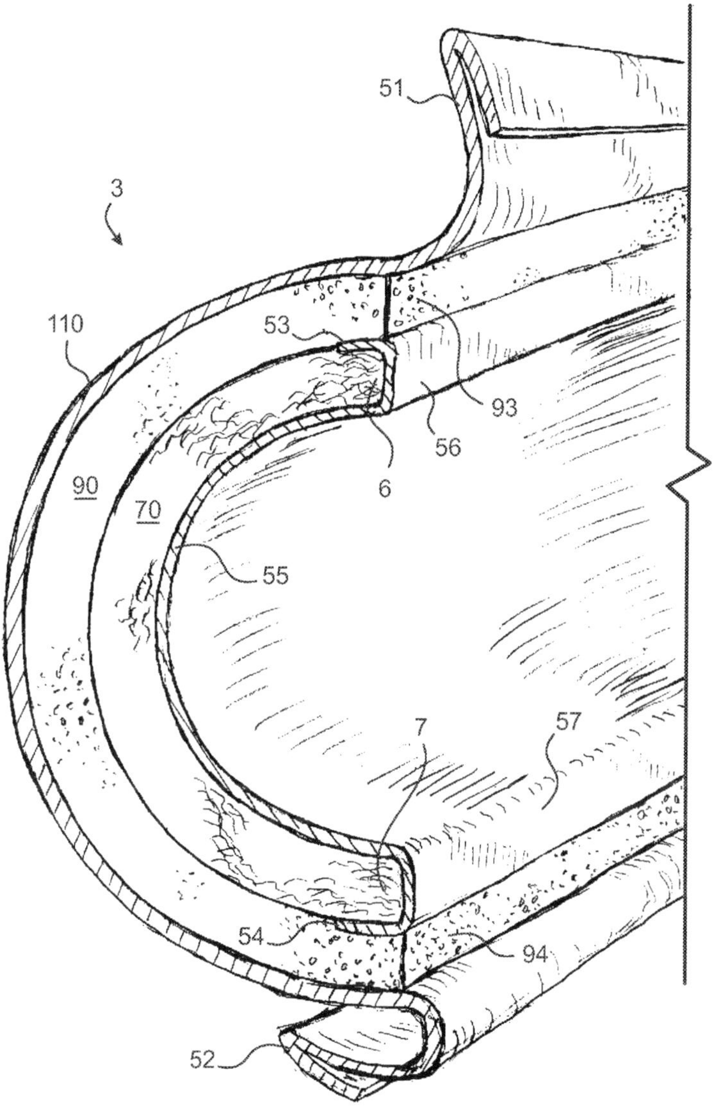

Figure 6.6 Heater element.

term outer is used to mean locations relatively further away from the gas line. When, for example, the gas line is a cylindrical pipe the term inner will generally refer to a radially proximal part of a cylindrically circumferential layer of material whereas outer will generally mean a radially distal part.

Furthermore, each cover piece 3 can have end portions, which remain exposed or abut other heater systems. These end portions benefit from protection when they are exposed. Generally, the end portions can be protected by radially extending end walls made from material similar to the wrap layer 110 or parts of the heater containing layer 55. Alternately, depending on the material used for the protective end walls, the end walls can be formed by dipping each of the axial ends of the heater cover pieces in liquified material which adheres to the ends and is then allowed to cure and solidify.

Figure 6.6 shows that the heater element containing layer 55 can have parts forming seam flaps 56,57 which extend radially over the axially extending seam portions 6 of each piece. In other words, the heater cover pieces 3 can mate to one another or to themselves along one or more axially extending seams, which can include portions which benefit from protection while the pieces are unmated. In other words, while a heater system cover piece is unmated, seam portions could otherwise be exposed without separate covering structures such as the seam flaps. Thus the thermal transition layer 70 can be encapsulated by parts of the heater containing layer 55, including seam and end flaps, and the silicone foam rubber insulation layer 90. In order to provide a more rugged structure to the seam flaps, each flap can have an extension 53,54 which doubles back upon the radial outer surface of the transition layer and underneath a seam abutting portion 93,94 of the silicone foam rubber insulation layer. The wrap layer 110 can have overlaps 51,52 which are shaped and dimensioned to cover the seams between cover pieces.

Preferably, the materials selected to form the various layers of the heater system are selected so that they do not adversely interact with adjacent materials.

A structural example of the materials and thicknesses of such a heater system is shown in Table 6.5.

A silicone foam rubber can be used as the insulating material on a heater which can be operated at much higher temperatures and for longer periods of time than earlier silicone foam rubber-based

Table 6.5 Components of heater system (32).

Layer	Material	Thickness/[[pn]]
Interface	Fiberglass impregnated silicone rubber – high temperature	0.022
Adhesive	High temperature silicone rubber	0.0075
Heater element	Inconel foil-type	0.001
Adhesive	High temperature silicone rubber	0.0075
Electrical isolation	Fiberglass impregnated silicone rubber – high temperature	0.022
Adhesive	High temperature silicone rubber	0.015
Thermal transition	Pyrogel brand insulation	0.2
Adhesive	Sheeted silicone rubber	0.015
Silicone foam	Non-high temperature silicone foam	0.2
Rubber insulation	Rubber	
Adhesive	Sheeted silicone rubber	0.005
Wrapping	Fiberglass impregnated silicone rubber	0.028

heater systems. Furthermore, inexpensive mechanical thermal fuses can be used in higher heat applications (32).

6.11 Cosmetic Compositions

6.11.1 Crosslinked Silicone Rubber Powder

Silicone rubber powders can be used by blending in cosmetic compositions for improving the feel of the cosmetic compositions (33). Silicone rubber powders are widely used as raw materials for cosmetic compositions, external use preparations, and quasi-drugs, whether the system is oily or aqueous. In particular, hydrosilylation reaction-crosslinkable silicone rubber powders using a hydrosilylation reaction catalyst including a platinum type metal are widely used as silicone rubber powders due to rapid curing, easily controlled reaction, and the ability to select various types of curing conditions.

In the common production methods, the platinum hydrosilylation reaction catalyst is incorporated into the surface or interior of the silicone rubber powder due to crosslinking, and thus remains in the silicone rubber powder product. The platinum metal in the silicone rubber powder can react with other cosmetic raw materials. This causes changes of the properties or odorization, and may cause discoloration by forming platinum black or metal complexes. It is thus desirable to lower the platinum metal content in the silicone rubber powder as much as possible (33).

A method to produce a hydrosilylation reaction-crosslinkable silicone rubber powder includes washing the hydrosilylation reaction-crosslinkable silicone rubber powder using an aqueous solution. The aqueous solution can be heated to a temperature of 30°C to 99°C. The aqueous solution contains at least one type of surfactant. The hydrosilylation reaction-crosslinkable silicone rubber powder produced via this method has a platinum metal content of less than 3.5 *ppm* by mass. The hydrosilylation reaction-crosslinkable silicone rubber powder can be used for various applications, including for use in a cosmetic composition, which can also include a cosmetic raw material.

An example of a preparation method for the silicone rubber powder runs as follows (33):

Preparation 6–5: First, 77.0 parts by mass of a dimethylvinylpolysiloxane, 21.5 parts by mass of a dimethylmethylhydrogen polysiloxane capped at both molecular terminals with trimethylsiloxy groups and an isopropyl alcohol solution of chloroplatinic acid (at an amount of 50 *ppm* in terms of mass units of the content of the platinum metal of the present composition) were stirred uniformly at 5°C to prepare an addition reaction curable silicone composition. Thereafter, this composition was dispersed in an aqueous solution at 25°C that was composed of 50 parts by mass of purified water and 1.5 parts by mass of poly(oxyethylene) isocetyl ether. After this mixture was made further uniformly emulsified using a colloid mill, 50 parts by mass of purified water was added for dilution to prepare the emulsion.

Thereafter, this emulsion was heated and cured to prepare a uniform aqueous suspension of silicone rubber particles. Thereafter, the suspension was dried in a hot air circulating oven at 150°C to prepare the silicone rubber powder. Particles of this silicone rubber powder were spherical. The average particle diameter was 6.2 μm, and the JIS A hardness was 60. Moreover, the coloration of the silicone rubber powder was around 0.28 (b*), and the platinum metal content was 7.2 *ppm* by mass.

This powder is then used further to obtain a material with a platinum metal content of 2.5 *ppm* by mass (33).

6.12 Silk Fibers

A study has demonstrated how the sensitive property of the constituent parts of silkworm silk fiber can be applied to the field of soft composite materials (34). This concerns the structural and intrinsic functional characteristics of silk versus the macroscopic property and the realization of functional composites.

Silk fiber-reinforced silicon rubber (SR) composites have been fabricated with different fiber lengths. The key structural features of silk made of stiff nanocrystals, including hydrogen bonded β-strands and β-sheet nanocrystals, when embedded in a softer matrix demonstrate that the tensile strength as well as the stiffness of the composites are higher than those measured for SR reinforced with synthetic polyether-ether-ketone fibers.

Moreover, the intrinsic luminescence of protein nanocrystals permits the direct observation of the deformation with accurate measurement of the strain in the composite.

The insertion of silkworm silk fibers in a stretchable rubbery matrix was found to be a powerful yet simple system for predicting the composite stiffness and probing the strain with luminescent emission of the silk patterns in the composite (34). The advantages of this system include the commercial availability of the fibers, silkworm silk, and the simple composite casting method by poly-addition of commercial silicone rubber. The degummed silk strands were found to be stiffer than monofilament poly(ether ether ketone) fibers. By using the luminescence pattern of silk fiber materials an optical strain gauge could be realized. It was shown that such a strain gauge can measure the deformation by using the intensity ratio of the emission spectra and the measurements reflected the material deformation measured by a universal tensile testing machine. It is expected that these results can be extended to obtain strain sensors to monitor the nonlinear constitutive laws of polymeric composites with real-time measurements. Moreover, these silk-based strain gauges would have the potential for *in-situ* strain measurements of soft tissues, which might be used to characterize the biomechanical function of biological tissue.

Based on the intrinsic properties of such natural hierarchical material, these findings will allow transferring the engineering of composite materials, particularly soft functional composites, to new applications ranging from strain measurements in biological tissue to monitoring tools in structural composites (34).

6.13 Elastic Silicone Rubber Belt

Circular knit fabrics may be required to undergo successive processing operations during their preparation, dyeing, finishing and making-up thereof, both in tubular form, as they descend from the circular knitting machine, and in open form, obtained by cutting the tubular fabric along one of its side edges (35).

Generally, before undergoing wet treatments such as, for example, dyeing, depending on the type or types of fiber which form the fabric, the fabric is subjected to a heat setting operation. In particular, heat setting is performed on fabrics composed of or partly containing synthetic fibers, such as thermoplastic fibers. The aim of heat setting is to fix the dimensions and the flat state of the surfaces of the fabric, thereby providing the fabric with stability. Heat setting a fabric also eliminates permanent creases or distortions that occur during the course of processing in bleaching and dyeing machines.

With common heat setting machines, it is not possible to differentiate between the processing of fabrics with an orthogonal pattern, i.e., of the weft/warp type, and the processing of knitted fabrics in general. This results in an increase in the duration of the processing cycle due to cutting and opening of the tubular fabric for heat setting and, sometimes, re-stitching of the fabric in a tubular form for the bleaching/dyeing operation, followed by reopening for the finishing and making-up operation. Moreover, these machines are unable to perform the heat setting of circular knitted fabrics, which must be completely processed in tubular form either for technical reasons, cost-related reasons, or because of market requirements (35).

Processing of fabrics in tubular form without heat setting results in considerable risks from the point of view of quality and is only performed in exceptional cases. In the past, various attempts have been made to develop specific machines for heat setting fabrics in tubular form. However, the results obtained have been somewhat unsatisfactory. The known machines are characterized essentially by the method of conveying and guiding the tubular fabric and by the system for transferring heat to the fabric.

In some machines, the tubular fabric is guided by means of a flat expansion device and is fed in a flattened form between the surface of a heated steel cylinder and an endless felt belt. The main drawbacks of these machines include the squashing of the side edges of the tubular fabric and a non-uniform heat exchange on its two surfaces. In other machines, the tubular fabric is guided by means of a flat expansion device which accompanies it in a flattened form through a horizontal or vertical chamber until it emerges therefrom. Inside the chamber, the two surfaces of the tubular fabric are acted on by flows of hot air. The heat exchange on the two surfaces of the fabric, however, is inadequate (35).

In a new textile manufacturing process, an elastic belt having a definite length is used. During the manufacturing process, in a section of the machine used to heat set the textile or cloth being produced, the definite length of the elastic belt elongates from 50% to 70% of its original length. Currently, rubber and nitrile rubber belts reinforced with yarns or fibers having a standard elasticity are used in the new manufacturing process. In addition, belts made only with rubber have also been used.

These types of belts, however, have exhibited a short service life when used in the new textile manufacturing process. The service life of these prior belts are shortened even more when subjected to the elevated temperatures in the heat setting sections of the textile manufacturing machine. Therefore, the temperatures that can be used in the manufacturing process are limited. Furthermore, the prior rubber belts experience excessive shortening in width in the cross-machine direction when elongated in the machine direction, which creates problems in the manufacturing process.

So, a need exists for a rubber belt that does not substantially shrink in width in the cross-machine direction while it elongates from 50% to 70% of its original length in the machine direction. Furthermore, a reinforced rubber belt is needed that has a low risk of delamination between the rubber portion and the reinforcing portion of the belt while elongating in the machine direction.

A silicone rubber belt has been developed that is reinforced with a structure comprised of elastic yarns or fibers (35). An elastic yarn or fiber is a yarn or fiber with an elongation at its breaking point of between 40% and 400% of its original length.

In an attempt to construct the desired reinforced rubber belt, a coating system was used. That is, an endless base fabric was constructed and coated with silicone rubber. In order to have good mechanical properties, the belt was constructed to have a double or triple layer base. This type of belt, however, created a structure having a high risk of delamination between the woven base and the silicone rubber coating, especially when the belt was stretched to a high elongation.

Preferably the material is constructed by wrapping elastic yarns or fibers constructed from materials, such as natural rubber, LYCRA® and ELASTAN®, onto a cylinder or mandrel. The circumference of the mandrel should be close to the overall length of the elastic belt being constructed. At least one layer of elastic yarns or fibers are wrapped onto the mandrel. The number of layers of elastic yarns depend on the desired resistance of the belt to elongation.

A single layer or a plurality of elastic yarn or fiber layers can be achieved by wrapping a single elastic yarn or fiber onto the mandrel in a continuous manner such that a single layer or a plurality of layers are formed with the elastic yarns or fibers in each layer tightly packed adjacent to one another. In addition, a plurality of elastic yarns or fibers can be wrapped onto the mandrel to form one or more layers of tightly packed elastic yarns or fibers.

Once the elastic yarns or fibers are wrapped onto or applied to the mandrel, a high viscosity silicone rubber is added on the top of the elastic yarn or fiber layers. Silicone rubber is added in order to attain the belt's required thickness and to provide a textile contacting surface (35). The textile contacting surface may be designed to impart a smooth or a textured finish to the textile being formed thereon.

When the addition of the silicone rubber is completed, the mandrel is heated in order to polymerize the silicone rubber. A rubber belt constructed in this manner does not substantially shrink in width in the cross-machine direction, while it elongates in the machine direction where it is capable of elongating from 50% to 70% of its original length. The completed belt may be used in textile finishing applications to finish, for example, textiles constructed of cotton, wool, and synthetic fibers, as well as blends of different fibers (35).

6.14 Recycling and Devulcanizing

A process for recycling and devulcanizing rubber has been described (36). In particular, the materials come from industrial manufacturing rejects or from articles which have reached the end of their useful life cycle, and originate from specialized collection centers or landfill sites.

The molecular chains of vulcanized rubber are typically linked by means of chemical bonds which evolve across bridges of sulfur atoms, which link them transversely. The vulcanized rubber therefore has a structure which is analogous to that of a thermosetting polymer, and therefore cannot be reused as such in the productive cycle.

Therefore, a recycling process has to make it possible to break the chemical bonds created by the sulfur in such a manner that the elastomeric material returns to the original physical/chemical state prior to the vulcanization treatment and is ready to be reused in the productive cycle.

The process for recycling and devulcanizing vulcanized rubber consists of the steps of (36):

1. Grinding the rubber in a mill in order to shred it into particles with a size substantially between 5 *mm* and 15 *mm*,
2. Introducing the particles in an initial section of a two-screw co-rotating extruder having a length 64 times or more the external diameter of the screws,
3. Mixing the particles within the extruder, which is kept at a temperature of between 35°C and 350°C, due to the screw rotation at a speed between 15 *rpm* and 300 *rpm*, and
4. Obtaining an extruded product formed by the devulcanized rubber.

This process is very flexible with regard to the starting materials to be treated and, in principle, makes it possible to devulcanize all types of vulcanized rubber (36).

6.15 Mobile Robots

A flat-type pneumatic actuator aimed at being applied to mobile robots working in narrow spaces or wearable devices has been developed (37). The actuator is configured with small diameter *McKibben Artificial Muscles* which show high flexibility and generate contraction force in an axial direction. By integrating the artificial muscles in a parallel form, the actuator can contract in the longitudinal direction and bend in any direction by changing the applied air pressure pattern to the artificial muscles.

The actuator was designed by nonlinear finite element model analysis (38).

The structure of the actuator has been developed and the fabrication process was established (37). Here, liquid silicone rubber is poured into molds and artificial muscles are placed in a mold.

The developed prototype actuator was configured with three artificial muscles of length 50 *mm*, width 5 *mm*, and a thickness of 2.6 *mm*. The bending motion with the minimum radius of curvature of 21.3 *mm* and contraction motion with the maximum displacement of 9.3 *mm* in the longitudinal direction were performed (37).

References

1. C.R. Moran and W.C. Michie, Sensor cable, US Patent 6 586 723, assigned to University of Strathclyde (Glasgow, GB), July 1, 2003.
2. D. Kweon, C. Park, and H. Byun, Method for manufacturing synthetic leather using different liquid silicone rubber coating solutions, US Patent 9 925 752, assigned to Jeongsan International Co., Ltd. (Busan, KR), March 27, 2018.
3. T. Tasaki and Y. Tsuji, Two-part curable liquid silicone rubber composition, US Patent Application 20 150 259 585, assigned to Dow Corning Toray Co. Ltd., September 17, 2015.
4. D. Wu, F. Xu, B. Sun, R. Fu, H. He, and K. Matyjaszewski, *Chemical Reviews*, Vol. 112, p. 3959, 2012.
5. V.A. Korzhikov, E.G. Vlakh, and T.B. Tennikova, *Polymer Science Series A*, Vol. 54, p. 585, August 2012.
6. H. Zhang and P.K. Shen, *Chemical Reviews*, Vol. 112, p. 2780, 2012.
7. J. Narasimhan and I. Papautsky, *Journal of Micromechanics and Microengineering*, Vol. 14, p. 96, October 2003.
8. S. Lee, D.H. Kang, H.N. Kim, and K. Suh, A bond less fabrication method for homogeneous polymer microchannel by capillary force lithography, in S. Verpoorte, ed., *Book of Abstracts*, pp. 458–460, Groningen, The Netherlands, 2012. 4th Intern. Conf. on Miniaturized Systems for Chemistry and Life Sciences, University of Groningen, The Netherlands.
9. M. Gaitan and L.E. Locascio, *Journal of Research of the National Institute of Standards and Technology*, Vol. 109, p. 335, June 2004.
10. S. Mukhopadhyay, S.S. Roy, A. Mathur, M. Tweedie, and J.A. McLaughlin, *Journal of Micromechanics and Microengineering*, Vol. 20, p. 055018, April 2010.
11. X. Wu, H.A. Reed, Y. Wang, L.F. Rhodes, E. Elce, R. Ravikiran, R.A. Shick, C.L. Henderson, S.A.B. Allen, and P.A. Kohl, *Journal of the Electrochemical Society*, Vol. 150, p. H205, 2003.
12. L.P. Yeo, S.H. Ng, Z. Wang, Z. Wang, and N.F. de Rooij, *Microelectronic Engineering*, Vol. 86, p. 933, 2009.
13. X.U. Zhao-Yu, *Guangxi & Development of Chemical Industry*, Vol. 2, p. 5, 2005.
14. Wikipedia contributors, Whisker (metallurgy) — Wikipedia, the free encyclopedia, https://en.wikipedia.org/w/index.php?title=Whisker_(metallurgy)&oldid=872441184, 2018. [Online; accessed 25-January-2019].
15. Wikipedia, Whisker (Kristallographie) — Wikipedia, die freie enzyklopädie, https://de.wikipedia.org/w/index.php?title=Whisker_(Kristallographie)&oldid=183073359, 2018. [Online; Stand 25. Januar 2019].

16. N. Zolotareva and V. Semenov, *Silicon*, Vol. 7, p. 89, April 2015.
17. N.D. McKay, Jr. and P.J. Williamson, Roller construction for detritus removal, US Patent 7 186 207, assigned to The Evercare Company (Alpharetta, GA), March 6, 2007.
18. S. Hamilton, Improved contact cleaning hand roller, WO Patent 2 012 025 900, assigned to ITW CS (UK) Ltd., March 01, 2012.
19. F. Achenbach, Adsorption of particles on a material containing a silicone polymer, US Patent Application 20 180 055 328, assigned to Wacker Chemie AG, Munich (DE), March 1, 2018.
20. H. Kigami and Y. Watanabe, Silicone adhesive composition, adhesive tape, and method for producing adhesive tape, US Patent Application 20 170 009 115, assigned to Nitto Denko Corporation, Osaka (JP), January 12, 2017.
21. T. Kaeser, D. Abegglen, A. Sarioglu, A. Kollep, and M. Bacchi, Liquid seal rubber applied to a capsule, US Patent 9 694 957, assigned to Nestec S.A. (Vevey, CH), July 4, 2017.
22. O. Fond, G. Lavanchy, and A. Yoakim, Obtaining fluid comestibles from substances contained in cartridges, US Patent 5 402 707, assigned to Nestec S.A. (Vevey, CH), April 4, 1995.
23. S. Yoshida and S. Hirabayashi, Silicone rubber composition for thermally conductive silicone-rubber development member, and thermally conductive silicone-rubber development member, US Patent Application 20 160 122 611, assigned to Shin-Etsu Chemical Co., Ltd., Tokyo (JP), May 5, 2016.
24. A. Kumar, A.A. Mollah, A.K. Keshri, M. Kumar, K. Singh, K.D.V.S. Rallabhandi, and R. Seelaboyina, *Industrial & Engineering Chemistry Research*, Vol. 55, p. 8751, 2016.
25. R.D. Corsaro and L.H. Sperling, *Sound and Vibration Damping with Polymers*, American Chemical Society, Washington, DC, 1990.
26. R. Barron, *Industrial Noise Control and Acoustics*, Marcel Dekker, New York, 2003.
27. L.L. Beranek and I.L. Ver, *Noise and Vibration Control Engineering: Principles and Applications*, John Wiley & Sons, Hoboken, NJ, 2nd edition, 2006.
28. J.P. Arenas and M.J. Crocker, *Sound & Vibration*, Vol. 44, p. 12, 2010.
29. M.E. Delany and E.N. Bazley, *Applied Acoustics*, Vol. 3, p. 105, 1970.
30. X. Sagartzazu, L. Hervella-Nieto, and J.M. Pagalday, *Archives of Computational Methods in Engineering*, Vol. 15, p. 311, 2008.
31. M.R.F. Kidner and C.H. Hansen, *International Journal of Acoustics and Vibration*, Vol. 13, p. 112, 2008.
32. J.S. Smith, C.C. Clausen, J.H. Kreisel, and T.J. Restis, High temperature gas line heater system including silicone foam rubber, US Patent 9 578 689, assigned to Durex International Corp. (Cary, IL), February 21, 2017.

33. M. Wakita, Low platinum content hydrosilylation reaction-crosslinked silicone rubber powder, US Patent 9 561 171, assigned to Dow Corning Toray Co., Ltd. (Tokyo, JP), February 7, 2017.
34. L. Valentini, S.B. Bon, L. Mussolin, and N.M. Pugno, *Composites Science and Technology*, Vol. 156, p. 254, 2018.
35. G. Cassarino, Elastic silicone rubber belt, US Patent 9 222 208, assigned to Albany International Corp. (Albany, NY), December 29, 2015.
36. G. Maris, Process for recycling and de-vulcanizing rubber, US Patent 9 527 978, assigned to Maris Frateilli S.p.A. (Rosta (Turin), IT), December 27, 2016.
37. Y. Toyama and S. Wakimoto, Development of a thin pneumatic rubber actuator generating 3-DOF motion, in *2016 IEEE International Conference on Robotics and Biomimetics, ROBIO 2016*, pp. 1215–1220. Institute of Electrical and Electronics Engineers Inc., 2 2017.
38. M. Yuzaki, K. Iwata, S. Wakimoto, and T. Kanda, Development of a thin pneumatic rubber actuator generating 3 DOF motion – design using FEM analysis and fabrication of the actuator, in *ACTUATOR 2018; 16th International Conference on New Actuators*, pp. 1–4, June 2018.

Index

Acronyms

Chemicals

Boldface numbers refer to Figures

General Index

Also of Interest

Check out these other books by the author published by Scrivener Publishing

Functional Synthetic Polymers
By Johannes Karl Fink
Published 2019. ISBN 978-1-119-59202-0

3D Industrial Printing with Polymers
By Johannes Karl Fink
Published 2018. ISBN 978-1-119-55526-1

Polymer Waste Management
By Johannes Karl Fink
Published 2018. ISBN 978-1-119-53608-6

Materials, Chemicals and Methods for Dental Applications
By Johannes Karl Fink
Published 2018. ISBN 978-1-119-51031-4

Fuel Cells, Solar Panels, and Storage Devices
Materials and Methods
By Johannes Karl Fink
Published 2018. ISBN 978-1-119-48010-5

Chemicals and Methods for Conservation and Restoration
Paintings, Textiles, Fossils, Wood, Stones, Metals, and Glass
By Johannes Karl Fink
Published 2017. ISBN 978-1-119-41824-5

Additives for High Performance Applications
Chemistry and Applications
By Johannes Karl Fink
Published 2017. ISBN 978-1-119-36361-3

Metallized and Magnetic Polymers
By Johannes Karl Fink
Published 2016. ISBN: 978-1-119-24232-1

Marine, Waterborne, and Water-Resistant Polymers
Chemistry and Applications
By Johannes Karl Fink
Published 2015. ISBN 978-1-119-018486-7

The Chemistry of Printing Inks and Their Electronics and Medical Applications
By Johannes Karl Fink
Published 2015. ISBN 978-1-119-04130-6

The Chemistry of Bio-based Polymers
By Johannes Karl Fink
Published 2014. ISBN 978-1-118-83725-2

Polymeric Sensors and Actuators
By Johannes Karl Fink
Published 2012. ISBN 978-1-118-41408-8

Handbook of Engineering and Specialty Thermoplastics
Part 1: Polyolefins and Styrenics
By Johannes Karl Fink
Published 2010. ISBN 978-0-470-62483-5

Handbook of Engineering and Specialty Thermoplastics
Part 2: Water Soluble Polymers
By Johannes Karl Fink
Published 2011. ISBN 978-1-118-06275-3

A Concise Introduction to Additives for Thermoplastic Polymers
by Johannes Karl Fink.
Published 2010. ISBN 978-0-470-60955-2

Printed and bound by CPI Group (UK) Ltd, Croydon, CR0 4YY
17/09/2026
14964244-0001